T0136980

Harnessing Performance Variability in Embedded
and High-performance Many/Multi-core Platforms

William Fornaciari • Dimitrios Soudris
Editors

Harnessing Performance Variability in Embedded and High-performance Many/Multi-core Platforms

A Cross-layer Approach

 Springer

Editors
William Fornaciari
DEIB – Dipartimento di Elettronica
Informazione e Bioingegneria
Politecnico di Milano
Milano, Italy

Dimitrios Soudris
School of Electrical & Comp. Engineering
Natl. Technical Univ. of Athens
Athens, Greece

ISBN 978-3-030-06336-8   ISBN 978-3-319-91962-1   (eBook)
https://doi.org/10.1007/978-3-319-91962-1

This Springer imprint is published by the registered company Springer Nature Switzerland AG
The registered company address is: Gewerbestrasse 11, 6330 Cham, Switzerland

Introduction

This book aims at providing a comprehensive view of the possible solutions to the problem of ensuring dependable performance, by collecting the best practices from the embedded and high-performance worlds. There exists a broad range of possible sources of performance variability, spanning from the hardware (silicon) aspects up to the design of the applications and the management of the resources. To worsen the scenario, it is no longer possible to underestimate the impact of power and thermal management and of process variation, especially in the perspective to deal with reliability and aging issues. The goal of this book is to share the experience matured during the HARPA project and to complement such information with background concepts and chapters, to create a scientific reference for both academic and leading-edge companies.

The HARPA project's goal has been to enable next-generation embedded and high-performance heterogeneous many-core processors to cost-effectively confront variations and yet provide dependable performance: correct functionality and timing guarantees throughout the expected lifetime of a platform under thermal, power, and energy constraints. The HARPA team is composed of industry and academic partners across Europe specialized in fields covering all the abstraction layers, from hardware to application level. One of the primary objectives of HARPA project is accomplished by means of an accurate yet efficient modeling of performance variability in scaled technologies. Capturing and simulating variability threats is an important step toward providing solutions in mitigating them.

The HARPA solution revolves around a cross-layer approach encompassing all pertinent actors in the system stack. At the core of HARPA, a middleware implements an innovative control engine that steers software/hardware knobs based on information from monitors strategically dispersed in the system. We have explored a large range of such knobs and monitors across different abstraction levels. This engine relies on technology models to identify and exploit various types of platform slack—such as performance, power, energy, temperature, lifetime, and structural (i.e., hardware)—to restore the timing guarantees and ensure expected lifetime in the presence of time-dependent variations. The challenge of dependable performance is to ensure timing correctness, whereas for high-performance applications it is paramount to ensure load balancing during parallel phases and fast execution. Both are achieved with a combination of a fast control engine with reaction times in the range of a few msec and an OS-level engine reacting slower. These concepts have been demonstrated in three different representative application case studies from the wireless, flood monitoring, and rockfall monitoring domains.

Among different variability phenomena, attention is given to the BTI effects, since they are expected to be the dominant ones. For this purpose, a proper estimation flow that models the dependence of variability at circuit and block level on the operating conditions (temperature and voltage), workload, and technology parameters has been designed and developed. Experimental results have shown running different application benchmarks of interest. The flow is based on commercial ASIC tools and in-house scripts: from the transistor-level representation of the circuit, and the transistor-level (real) workload, BTI variation based on defect-centric models is quantified, and its impact on the critical path delay of the reference circuit is then analyzed through standard STA run.

Although the impact on BTI is driven by the trends in current technology, the system-level perspective is a joint result of different failure mechanisms. To close the gap with system-level analysis, an experimental setup to measure the impacts of stress-induced variability at the system level, from a functional perspective, has been designed and deployed. The proposed setup allows us to directly measure from the hardware the time between successive faults. While modeling and deeply understanding variability impact is an important part of the work, a fundamental objective of the project is to provide solutions to mitigate reliability threats and ensure dependable system performance. Toward this direction, the HARPA engine has been developed, implementing various control frameworks across the system stack. The goal is to exploit different manifestations of platform slack (i.e., slack in performance, power, energy, temperature, lifetime, and structures/components), in order to ascertain timing guarantees throughout the lifetime of the device.

The reader is steered into the concept developed during the project passing through three main stages: background concepts, design methodologies, and analysis of use cases. To simplify the navigation of the book to the reader, the discussion is further split into 6 parts and finally 13 agile chapters.

The beginning of the first part is an introductory chapter providing an overview of the target problems and of the approach to their mitigation developed during the HARPA project, which is a sort of reference guide to move into the book concepts and chapters. The other two chapters of the first part provide a background on the evolution of the computing platforms, with a particular focus on the processor architectures and some foundations on the aging effects from the physics of the phenomenon up to the CAD aspects.

The second part starts entering into the run-time management technologies with two chapters explaining first of all the solutions working at the operating systems level and then how to cope with the problem to achieve a fast and low overhead thermal control of multi-many core processors.

The following two chapters, constituting the third part, move into a finer grain of the design considering the RT level and how to design an ad hoc run-time engine to deal with real-time performance requirements. A comprehensive view of the monitors and knobs is the objective of the fourth part of the book, focusing on both the system-level solutions and on a specific new brick of the emerging platforms, namely, the Network-on-Chip. A coordinated application of the background and concepts presented so far is carried out in the two chapters composing the fifth part of the book. The first of them is affording the problem of properly modeling and simulating BTI-induced degradation in the computing platform and its impact on the timing performance, while the second is the description of a proof of concept performed on a real platform.

The final part of the book figures out how the methodologies and techniques here presented can be effectively applied on real-size problems. To this purpose, two use cases out of the three developed during the HARPA project are presented: Floreon+ to show the possible benefits achievable on application and architectures belonging to the high-performance computing domain and Beesper, which is a typical low-end embedded system submitted to severe reliability and energy management constraints.

This book is probably the first one trying to cover all the critical aspects of the computing continuum and, as we said, is well suited both for engineers working in leading-edge companies and for courses at the master's and PhD level.

Special thanks to Antonio Gonzalez and Jörg Henkel for their availability to cooperate with us in this project, writing two excellent chapters of the book and with the hope to extend in the future the scientific cooperation on new challenges.

We hope you will enjoy reading the book, as the authors enthusiastically worked together for 4 years.

Milano, Italy   William Fornaciari
Athens, Greece   Dimitrios Soudris

Contents

Chapter 1
The HARPA Approach to Ensure Dependable Performance

Nikolaos Zompakis, Michail Noltsis, Panagiota Nikolaou,
Panayiotis Englezakis, Zacharias Hadjilambrou, Lorena Ndreu,
Giuseppe Massari, Simone Libutti, Antoni Portero, Federico Sassi,
Alessandro Bacchini, Chrysostomos Nicopoulos, Yiannakis Sazeides,
Radim Vavrik, Martin Golasowski, Jiri Sevcik, Stepan Kuchar, Vit Vondrak,
Fritsch Agnes, Hans Cappelle, Francky Catthoor, William Fornaciari,
and Dimitrios Soudris

1.1 Introduction

The evolution towards processors with heterogeneous many-cores is impeded by variability related challenges. This is primarily due to increasing susceptibility to static [3] and dynamic [4] variations in the performance and leakage energy of the silicon devices and wires. Static variations are due to imperfections in the fabrication process, such as line-edge roughness and random dopant fluctuations, whereas dynamic ones are due to time-dependent variations, such as aging, ambient

N. Zompakis · M. Noltsis · D. Soudris
MicroLab-ECE-NTUA, Athens, Greece
e-mail: nzompaki@microlab.ntua.gr; mnoltsis@microlab.ntua.gr; dsoudris@microlab.ntua.gr

P. Nikolaou · P. Englezakis (✉) · Z. Hadjilambrou · L. Ndreu · C. Nicopoulos · Y. Sazeides
University of Cyprus, Nicosia, Cyprus
e-mail: panagiota.nikolaou@cs.ucy.ac.cy; englezakis.panayiotis@ucy.ac.cy;
zhadji01@cs.ucy.ac.cy; lndreu01@cs.ucy.ac.cy; nicopoulos@ucy.ac.cy; yanos@cs.ucy.ac.cy

G. Massari · S. Libutti · W. Fornaciari
DEIB Politecnico di Milano, Milano, Italy
e-mail: giuseppe.massari@polimi.it; simone.libutti@polimi.it; william.fornaciari@polimi.it

A. Portero · R. Vavrik · M. Golasowski · J. Sevcik · S. Kuchar · V. Vondrak
IT4Innovations, Ostrava - Poruba, The Czech Republic
e-mail: antonio.portero@vsb.cz; radim.vavrik@vsb.cz; martin.golasowski@vsb.cz;
jiri.sevcik@vsb.cz; stepan.kuchar@vsb.cz; vit.vondrak@vsb.cz

F. Sassi · A. Bacchini
Camlin Italy s.r.l., Parma, PR, Italy
e-mail: f.sassi@camlintechnologies.com; a.bacchini@camlintechnologies.com

© Springer International Publishing AG, part of Springer Nature 2019
W. Fornaciari, D. Soudris (eds.), *Harnessing Performance Variability*
in Embedded and High-performance Many/Multi-core Platforms,
https://doi.org/10.1007/978-3-319-91962-1_1

conditions, and dynamic workload properties [7]. Variations, therefore, introduce an unintended heterogeneity, as compared to design time heterogeneity, that needs to be dealt with.

Application requirements, as well as power and technological constraints, are driving the architectural convergence of future processors towards heterogeneous many-cores. This development is confronted with variability challenges, mainly the growing susceptibility to time-dependent variations in silicon devices and wires. A variety of well-known mitigation approaches, such as tuning manufacturing process, column/row sparing [9], frequency binning and body biasing [3],dynamic sparing [10], variation tolerant circuits [8], and lifetime guard-bands [1], have been adopted to provide chips a number of solutions for mitigation techniques for such variability issues. The aforementioned solutions provide a basic but key expectation for a reliable processor that throughout its lifetime performs correctly under any nominal operating condition. The goal of the deployed HARPA methodology is to enable next-generation embedded and high-performance heterogeneous many-core processors to cost-effectively confront variations and yet provide dependable performance: correct functionality and timing guarantees throughout the expected lifetime of a platform under thermal, power, and energy constraints. The proposed HARPA solution revolves around a cross-layer approach encompassing all pertinent actors in the system stack. A middleware implements an innovative control engine that steers software/hardware knobs based on information from monitors strategically dispersed in the system. This engine relies on technology models to identify and exploit various types of platform slack—such as performance, power, energy, temperature, lifetime, and structural (i.e., hardware)—to restore the timing guarantees and ensure expected lifetime in the presence of time-dependent variations. The challenge of dependable performance is critical for embedded applications to provide timing correctness, whereas for high-performance applications it is paramount to ensure load balancing during parallel phases and fast execution during sequential phases. The lifetime requirement of dependable performance is crucial, with ramifications on the cost of tuning manufacturing process and the number of field returns (products returned to the manufacturer due to faults). The novelty lies in the seeking of synergies in techniques that have been considered virtually exclusively either in the embedded or high-performance domains (e.g., worst-case guaranteed partly proactive techniques in embedded and dynamic best-effort reactive techniques in high performance). This chapter outlines the merging concepts from the two domains of embedded and high-performance platforms by highlighting the key issues from both segments.

F. Agnes
Thales Communications & Security, Gennevilliers Cedex, France
e-mail: fritsch.agnes@thalesgroup.com

H. Cappelle · F. Catthoor
Imec, Leuven, Belgium
e-mail: hans.cappelle@imec.be; catthoor@imec.be

A comprehensive framework (along with its associated methodology and supporting tools) ensures performance dependability in the many-core systems of the near future. Specifically, the Engine—the core of the proposed framework—has direct applicability in the existing systems, and provides extensive architectural support for future SoC platforms. The deployed components have been validated using various important use cases, which have been selected due to their significance in emerging application trends.

1.2   Dependable Performance Through Guaranteed Quality of Service

The means to achieve dependable performance is to enrich future processor architectures with capabilities that provide nonfunctional guarantees, such as timing, power, lifetime, etc., in the presence of time-dependent variations. In other words, we want to provide a *quality of service* (QoS) for specified nonfunctional requirements while transistors and wires delay and leakage is changing with time as presented in the example of Fig. 1.1.

Some primitive notions of HARPA principles are present in the existing systems (such as load balancing and dynamic voltage scaling, etc.) but are not pervasive enough and not orchestrated or meant to provide dependable performance. Another unique aspect of HARPA is in seeking synergies in techniques that have

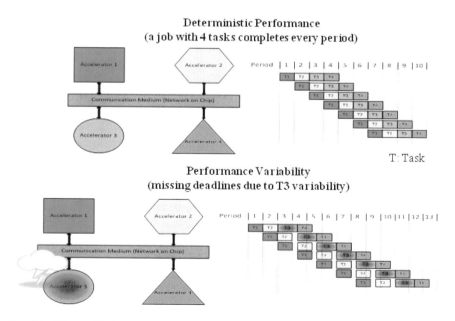

Fig. 1.1 Effects of PV execution of a periodic application with four tasks each on a different accelerator (above) Free of PV (below) PV due to the Accelerator 3 results in missing deadlines

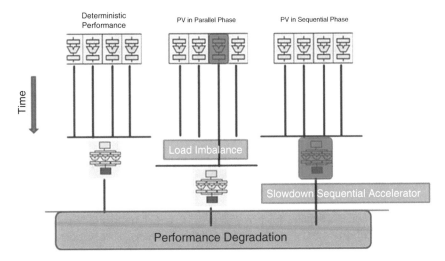

Fig. 1.2 Execution modes and performance degradation—Execution of a program with parallel and sequential phases: (left) PV free, (middle) PV in one of the cores used for parallel execution results in load imbalance and degradation, and (right) PV in the core used for sequential execution results in degradation

been considered virtually exclusively either in the embedded or high-performance domains, such as proactive techniques in embedded and reactive techniques in high performance.

Overall, the aim is to deploy the necessary architectural and micro-architectural, middleware and system support as well as the cross-layer interfaces to express and efficiently provide the following guarantees: timing, power (peak and energy), lifetime, and temperature in the presence of time-dependent variations.

While instances of such concepts have been widely adopted in the network community, surprisingly they are only now becoming a serious issue also in the converging domains of high-performance and embedded computing systems. Due to the different characteristics, the earlier solutions cannot just be copied: significant adaptations and extensions are required. The dependable-performance concepts are becoming more than ever relevant because of the increasing complexity and dynamism that is present in the application workloads, where several tasks with highly diverse requirements are running concurrently on the systems, as well as the dynamicity of the hardware due to time-dependent variations. This is clearly articulated in the HiPEAC roadmap as a need for dependable timing and predictable performance [5] as presented in the example of Fig. 1.2.

The aim is therefore to augment embedded and high-performance platforms with mechanisms to:

- Provide dependable performance depending on the specific nonfunctional requirements and actual load of the system.
- Expose nonfunctional time-dependent hardware state (such as delay, power, temperature, and lifetime) across compute layers to enable proactive and reactive control of a platform.

- Shave unnecessary margins to avoid over-provisioning or underutilizing the architecture.

1.3 Concept and Principles

Designing either embedded systems or high-performance, general-purpose processors with the innate capability to provide dependable-performance guarantees requires a clear separation of control *policies* and *mechanisms*. *Policies* provide solutions; for flexibility, we strive for policies implemented in software, and, in particular, in the system software layer so that the application designer does not need to worry about them. *Mechanisms* provide the primitives for constructing policies. Examples of such mechanisms in systems are task migration, dynamic voltage and frequency scaling, power gating, component disabling, etc. Because primitives are universal, system designers can implement mechanisms in both hardware and software. In general, mechanisms that interact directly with fine-grain hardware resources should be implemented in hardware; to reduce hardware cost, mechanisms that manage coarse-grain resources should be implemented in software. The key issue is to provide integrated *policies* and *mechanisms* in a cross-layer implementation that holistically aims to provide dependable performance in the presence of time-dependent variations and aging.

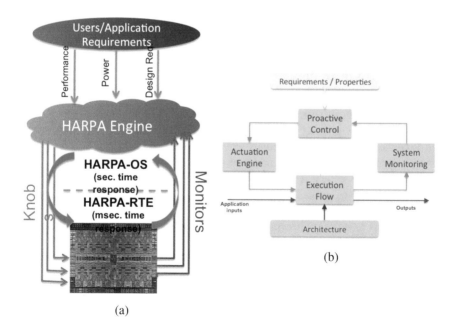

Fig. 1.3 HARPA framework. (**a**) HARPA high-level view. (**b**) Conceptual view

Figure 1.3 below shows the main concepts of the proposed architecture and the main components that can provide performance-dependability guarantees. Note that this generic framework applies to both embedded systems and high-performance general-purpose systems. We outline below the three main elements that distinguish: *Monitors and knobs* to implement mechanisms that can provide and maintain performance-dependability guarantees, and the *HARPA Engine*.

Conceptually, the latter mainly consists of a feedback loop (See Fig. 1.3b), where the different metrics (performance, timing, power, temperature, errors and manifestations of time-dependent variations, etc.) of the system are continuously monitored. The HARPA engine actuates the knobs to bias the execution flow as desired, based on the state of the system and the performance (timing/throughput) requirements of the application. It is the HARPA engine that will implement the various control strategies aiming to provide dependable performance in the presence of (highly) unreliable time-dependent variations. The goal is to exploit different manifestations of what we term as platform slack (i.e., slack in performance, power, energy, temperature, lifetime, and structures/components), in order to ascertain timing guarantees throughout the lifetime of the device (in spite of time-dependent variability) and maintain the expected lifetime of the system. By combining performance dependability techniques from both the embedded systems and high-performance worlds, HARPA has enabled cross-fertilization of techniques and mechanisms from these two converging domains and transfer methodologies from one to the other. The underlying architecture is a heterogeneous multicore architecture, which consists of a single-ISA multicore environment, where cores have different performance and power characteristics (low-power slower cores, high-performance fast cores, and application-specific accelerators).

1.4   User Requirements

Because performance-dependability guarantees are often application specific, the HARPA approach provides an efficient and general interface that can satisfy performance-dependability objectives over a range of applications. In this chapter, we will investigate several different application domains of the HARPA engine, and define the necessary API and language support to express the requirements. The applications to be analyzed cover both the embedded systems domain and the high-performance computing (HPC) domain. A common factor for all type of applications is that the application requirements provided to the HARPA architecture have to be architecture-independent. This removes the need for the application programmer to provide different sets of metrics for different target architectures, and this attribute will guarantee performance and requirements portability. In particular, the HARPA engine will be able to handle the disparate, yet converging, demands of embedded systems and HPC for timing predictability and performance determinism. In the case of some mission-critical embedded applications, one needs much stronger guarantees and a pure best-effort approach is not sufficient. In that case, an extended analysis phase is required to ensure that the HARPA engine will be able to provide such strict guarantees. On the HPC side, lack of precise knowledge of the applications (due to the fact that high-performance systems are expected to run a variety of applications whose control flow can be data sensitive) necessitates dynamic profiling and a combination of proactive and reactive (post-failure) corrective measures.

1.5 HARPA Engine

The HARPA engine is the heart of the proposed methodology. Its role is to *proactively* and *reactively* control the system resources, in order to guarantee the performance requirements of the various applications running on the heterogeneous architecture, establish the necessary service-level agreements (SLA) with the system, periodically monitor its state, and select and steer the appropriate knobs to provide the necessary performance guarantees throughout the lifetime of the system and in the presence of time-dependent variations and aging. It should be noted that the notion of SLA is somewhat different for embedded systems as opposed to HPC systems. In the former case, SLA primarily focuses on satisfying certain constraints (meeting deadlines), while in the latter case, it implies the elimination or minimization of any deviations from the manufacturer's nominal specifications (e.g., cache capacity, core performance, and availability, etc.). With this subtle difference in mind, we address both challenges through the HARPA engine, which is a middleware split between the Operating System (HARPA-OS) and the Run-Time system (HARPA-RT), as shown in Fig. 1.4.

This division is done with a two-fold objective: obtain better semantic information from the application, and enable high dynamicity. The run-time layer (HARPA-RT) sits at a low level in the system stack and it is in direct contact with the various monitors and knobs. It is an application-independent layer (even though it adapts to application-specific semantics during execution) that has direct and responsive control on hardware resources. Its low-level nature

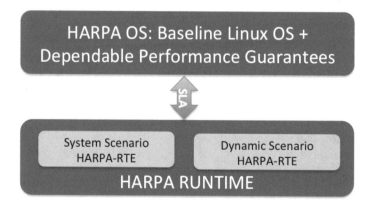

Fig. 1.4 HARPA engines layers and types

(directly controlling/actuating hardware resources) allows the run-time engine to respond at a *millisecond time granularity*, thus enabling extremely fast adaptation to system behavior, which is ideal for providing guarantees for the hard real-time specifications of an application. Essentially, the HARPA-RT layer is a type of "firmware" running on top of (or beside) the actual computational resources and is under the overall control of the other major component of the HARPA engine, which operates at the overall Operating System level. The HARPA-RT alone is not enough, because when multiple applications are running concurrently, each active run-time application cannot have a holistic, global view of the system. To ensure system-level performance dependability, a component of the HARPA engine must operate at the Operating System level. This will allow for system-wide guarantees. The OS is expected to maintain a piece of software (e.g., a library) linked with each active application, which provides a set of common services to support its run-time optimization. Note that this OS library is "by construction" semantically closer to the running applications, which allows it to obtain salient semantic information about the run-time applications (such as the phases of the application and its behavior). Thus, this library supports the smooth and proper integration of a run-time application with the HARPA-OS. Naturally, the HARPA-OS component is slower to react than the run-time component (HARPA-RT), as the OS's operation involves the invocation of system calls to/from the active run-time applications, or similar mechanisms that require changing from user to kernel modes. This implies that the OS reacts/adapts to system changes at the *second time granularity*. However, this is perfectly adequate for less-critical, best-effort performance requirements, and nonfunctional requirements that evolve at coarser time granularity, such as aging and overall power/energy and thermal management. By splitting the HARPA engine into both layers (*millisecond-scale* run-time and *second-scale* OS), we can get the best of both worlds. The fast HARPA-RT layer cooperates with the coarse-time granularity of the HARPA-OS layer to achieve system-wide optimization goals by still granting a prompt low-level control on hardware resources. Two different HARPA Run-Time approaches are investigated: (1) System Scenario control, and (2) Dynamic Scenario control. Both are complementary and most applicable for different types of applications and system context.

HARPA is tasked with the responsibility of ensuring that performance variability remains bounded, in order to guarantee certain predefined performance dependability levels. The amount of performance variability that can be afforded in a system depends on the scenario: in an embedded system, PV may be tolerated up to the point where timing constraints are no longer met (deadlines are lost), whereas in an HPC setup, the system may only tolerate performance degradation up to a specified level within a given amount of time. Thus, HARPA must ensure that the combined monitor/control/knob reaction latency is never exposed to the application, or it is only exposed within a given bound.

The operation of the HARPA engine is summarized as follows:

- *Application setup and characterization*: As mentioned earlier, the application is first augmented with the user requirements. At installation, the run-time engine is configured with the characteristics of the hardware (e.g., the sizes of the caches and the processor core functionality, but also the symbolic models of the expected degradation mechanisms). Additionally, in the case of application-specific embedded systems, the run-time engine may also be configured with the application characteristics (e.g., the nature of the operations and control flow that will be executed in the started application task set, the type of data access patterns, etc.). Application characteristics are obtained by profiling, or by any other available means of application analysis. This will not require internal information of the application, but it will suffice to have access to executable code, or a hardware emulation board that can be supplied by the application architect or the application designers. This will heavily reduce the design-time overhead and it will remain one of the main objectives to keep it small. On the other hand, in the case of high-performance systems, the applications may not be precisely known a priori.

- *HARPA Engine*: The role of the HARPA engine is to observe the application behavior at run time (dynamically), in order to extract insight as to the application's characteristics. If multiple different applications are running concurrently, then the HARPA engine will observe the combined/aggregate behavior. In such a situation, where not enough information is available for the HARPA-OS to manage resources, the architecture is required to remain in an optimal state with respect to high-level objectives that may be provided as directives from the HARPA-OS. In this particular scenario, the hardware can see all the running applications as a generic workload on top of the resources, since no user/application requirements have been provided. Moreover, these hardware optimizations are required to avoid fine-grain and hardware-dependent focus for the HARPA-OS. In particular, there are at least three possible scenarios. First, a non-profiled set of applications can run on the HPC platform. To this extent, while the HARPA-OS collects behavior information it can steer the low-level optimization layer (eventually hardware based) to a particular power-performance trade-off configuration. After a while, the HARPA-OS can smoothly start taking control of the resource management, while it has acquired enough information on the application behaviors. Second, the low-level/hardware optimization layer is required to manage exceeding resources not required by the HARPA-OS following the policy stated above, i.e., high-level directive to optimize towards a specific power-performance or reliability-performance trade-off. Last, it is possible to have applications with a very fast dynamic or with operating modes that changes too fast, thus they cannot be efficiently managed by the HARPA-OS. In that case, the RT engine should still be able to deal with this part of the dynamics (up to msec-level updates). Only when the dynamics are even faster, an averaging effect will have to be allowed and this will reduce the cost efficiency. Still, even then the dynamic scenario approach will still allow to guarantee the hard real-time constraints, albeit with a higher cost due to the "gas pedal" concept it requires to maintain that guarantee.

- *Service-Level Agreement (SLA)*: Before any new application task set starts, the HARPA engine verifies whether the performance requirements can be met with the current hardware and middleware status (e.g., are enough memory and processor resources available given the other tasks that are already active and committed). The SLA is then validated. If not, the HARPA engine will report the SLA invalidation to the system-level controller. On top of this, in the proactive mode with the System and Dynamic run-time approaches, the HARPA engine also analyzes whether the expected evolution (including degradation and dynamic behavior) of the hardware and system software allows for the satisfaction of the desired performance requirements. This analysis will be based on predicted-but-safe bounds that are produced and updated at run time. How this is achieved will depend on the nature of the HARPA engine mechanisms at the run-time system and OS layers. If this check is positive, the task can start and the performance requirements will be met until the task's predefined ending time: the SLA is then guaranteed and, hence, fully validated. If the check fails, then the HARPA engine will report this to the system-level controller, which has to initialize a backup plan for the task. For instance, in a wireless system, the data transmission of a particular user can be postponed till later, or another route can be employed to deliver it. In a mission-critical airplane take-off or car/train ride, the pilot or driver will decide not to start and will proceed to the appropriate recovery procedures, rather than risk a failure during the flight/ride. This is fully compatible with the current procedures for such applications. In the case of HPC setups, compliance with the SLA is monitored and assessed dynamically to ensure that all running applications experience the expected performance determinism and the system's expected lifetime guarantee is not jeopardized. Any deviation from these conditions that cannot be corrected by the HARPA engine will be reported to the user/administrator layer.

• *Execution-time behavior of the HARPA engine*: At execution time, the HARPA engine will adapt to the prevailing run-time conditions, and "based on the difference between the predicted bounds and the actually required resources" it will compute the available slack. That slack will be used to shave over-provisioning margins as much as possible. When the SLA cannot be strictly met at task set start-up (see above), we will perform a best-effort approach (e.g., a 90% guarantee instead of 100%), once the user has been notified and has accepted this downgrade in performance guarantees. When the assumptions are fulfilled, the HARPA engine will meet the performance guarantees during the task execution and it will also account for application tasks with irregular and data-dependent control flow (including conditions and while loops). In this way, we will be able to reduce the cost much more than any worst-case bounding that happens at design time. Moreover, in contrast to best-effort approaches, we will continuously adjust the HARPA knob control to meet the performance requirements, once the initial check at task start-up has given the go-ahead (in embedded systems), and maintain performance determinism and lifetime guarantees (in HPC systems). The ability to identify the impact of time-dependent variations and aging at run-time, and to dynamically adjust system behavior to account for such attrition, is one of our main differentiators. The HARPA engine will proactively strive to maintain performance dependability, and it will reactively address any permanent failures that happen as a result of aging, with the ultimate goal being that the user still receives the promised performance guarantees.

1.5.1 HARPA Run-Time Engine Implementations (HARPA-RT)

The run-time system is the low-level component of the HARPA engine and it directly controls the hardware monitors and knobs. By directly actuating the hardware resources, the HARPA-RT can provide extremely fast adaptation to application behavior at the *millisecond* time granularity. Even though the HARPA-RT layer itself is application-independent, during execution it receives application-specific semantics from the HARPA-OS (through the application-linked library previously described). This interface with the OS allows the HARPA-RT to be semantically in direct contact with the application during execution. Two key aspects in this interaction are (1) how dynamic is the behavior of the application, and (2) the required information by the run-time mechanism to track/predict the future resources required by the application to meet its performance requirements. In this chapter, two different run-time technologies will be investigated: the "System Scenario" and the "Dynamic Scenario" approaches. The characteristics and use of each technology are mainly steered by the dynamicity on the application. Note here that even though different platforms will inevitably feature different degrees of observability and controllability, the *interface definitions* that have been developed (between the platform's hardware and the HARPA-RT) will be agnostic to the specifics of the underlying platform, in order to enable widespread applicability. Hence, even though the values of the parameters passing through the interface will be platform specific, the interface definition itself will be platform-independent.

1.5.1.1 System Scenario Run-Time Engine

The System Scenario [6] approach may be defined as an approach where the characteristics of the scenario in which a system operates are exploited in a specific way to reduce margins. The System Scenario approach is typically suitable for mildly dynamic applications (due to conditions and while loops) or static applications. Based on a specific grouping of the possible application behaviors that can occur at run time, a clustering is made into the so called system scenarios. For example, in the case of an MPEG4 video codec, the code and note can be profiled, while all of the frames of a given type have similar E(nergy)–T(ime) relation, so they have a stable behavior that can be tracked/predicted safely by the run-time mechanism to meet all performance requirements of the application. However, other frames of a given type P internally may have a too wide distribution for the E–T curves. In that case, P frames could be further broken down into sub-frames, whose behavior should remain relatively stable across the active lifetime of the application. Obviously, this class of applications includes also the purely static applications where no data-dependent behavior is present, such as in pure streaming applications. To carry out this clustering/splitting of the application into sub-tasks, the run-time engine needs profiling/semantic information about the frequency of occurrence and the distance between sub-tasks. These will be carefully characterized at design time (Application Setup phase) in terms of their spatial and temporal requirements on the platform. In the HPC case, such static analysis may not always be feasible (or may only be partial), which transfers the burden to the dynamic run-time engine.

The main advantage of the System Scenario approach is that if the different System scenarios are fully characterized and bounded, and if the OS layer honors the SLA agreement (availability of resources), then a strict guarantee can be provided on critical requirements, such as timing, for example. This is very useful in mission-critical applications with hard real-time constraints. In addition, unlike conventional techniques where the WCET of the application has to be fully determined upfront, the engine is able to adapt to systems change (detected by the monitors), and yet provide strict guarantees as long as the assumption of each System scenario is fully characterized. More details on how the System Scenario approach is operating can be found in [6].

1.5.1.2   Dynamic Scenario Run-Time Engine

The Dynamic Scenario approach, initially conceptualized in [11], mainly extends System Scenario techniques for highly dynamic applications (such as 5G wireless radio paradigms, scalable multimedia codecs, or smart adaptive closed-loop biomedical algorithms) and takes into account the presence of strong nonlinearities that are present in the system, such as the mapping between the data and the hardware memory organization, or the relation between the temperature and the delay and energy that are required to execute the application tasks. The HARPA run-time system has to proactively deal with the expected changes in the near future and the nonlinearity makes these predictions highly complex. To achieve this, the dynamic scenario approach exploits correlations which are present in the run-time events of any realistic application context. That is one of the most demanding issues that has been addressed in the current approach. The challenge is how the design goals can be achieved reducing cost (further shaving of margins) while guaranteeing all constraints. That will definitely require pushing more of the analysis and decision-making to the execution phase of the application. Moreover, it is needed to increase the look-ahead capability by introducing safe bounds that are not based on upfront design-time bounding. This aspect is even more relevant in HPC setups, where inherent dynamism in the application set necessitates dynamic profiling and run-time observations in order to actuate proactive and reactive mechanisms. If the proactive efforts fail to avoid a permanent failure (occurring due to aging and time-dependent variations), the HARPA run-time engine will resort to reactive techniques that will aim to maintain performance guarantees (e.g.,

Table 1.1 HARPA run-time engine summary

Approach	Characteristics	Examples of applications	
		HPC context	ES context
System scenario	Static or mildly dynamic applications	Graphics rendering	MPEG video
	Assume predictable resource availability		
	Requires upfront characterization of the application by profiling (gray box)	Reduced number of executable paths	Observability of application control flow (i.e., I or P frame types)
	Can provide strict guarantees under certain conditions		
Dynamic scenario	Highly dynamic application	Infrastructure as a service (IaaS)	5G wireless services
	Subject to nonlinear dynamic system Behavior		
	Requires some characterization of the application behavior (gray box)]	Stochastic resource requests and virtualization overheads	Exploding the number of possible execution paths; reduced scenario detection overhead
	Can provide strict guarantees under certain conditions		

through some reconfiguration, component deactivation, remapping, etc.) despite the decommissioning of some elements. Table 1.1 summarizes both HARPA engine approaches.

1.6   HARPA Operating System (HARPA-OS)

The Operating System is the second layer of the HARPA engine, as previously depicted in Fig. 1.3a. For its implementation, we started from the works introduced in [2] and [13]. The OS offers the advantage of a global view of all system requests and resources, which enables system-wide performance-dependability guarantees. Its main limitation is that it is slower to react than the HARPA-RT component, providing services at the time granularity of a second. Regardless, such granularity is adequate for global system-level coordination, and performance-dependability guarantees in the presence of multiple concurrently running applications. Following the HARPA philosophy of seamlessly ensuring dependable performance, our aim is to enable a mainstream operating system (OS) with such system-level capabilities.

The OS has to offer some guarantees in the access to system resources to the different run-time applications running on top of it, so that the HARPA-RT layer can offer guarantees in accomplishing user requirements. The OS also has to take into account the constraints imposed by the system administrator, such as a thermal threshold or a power cap. One objective of the HARPA engine at the OS level, HARPA-OS, is that it does not deny too quickly (or conservatively) the SLAs and that it makes all the optimizations necessary to grant the SLA upfront.

The HARPA approach is that the HARPA-OS will provide a set of virtualized resources to each run-time application, making each of them think as if they are alone in the machine. At run time, when each application starts, it sets an SLA with the OS in which it specifies the resources needed. At that moment, the OS maps the virtualized resources into actual resources by spatially and timely splitting actual resources. At low level, The HARPA-OS guides the HARPA-RT's direct interaction with the different hardware/monitors provided by the architecture, as described below.

1.6.1 Resource Virtualization

In a traditional multiprogrammed system, the OS assigns each application a portion of the physical resources for example, physical memory and processor time slices. From the application's perspective, each application has its own private machine with a corresponding amount of physical memory and processing capabilities. With multicore systems containing shared micro-architecture level resources, however, an application's machine is no longer private: the resource usage of other independent applications can affect its resources. The HARPA engine virtualizes the hardware resources to the applications/run-time system. At run time, these virtual resources will be mapped to actual hardware resources. This provides the user with a single interface, regardless of the underlying hardware, removing the need of adapting the application to different target architectures. This virtualization provides flexibility to the HARPA engine when dealing with different applications. During the SLA establishment, the run-time system will negotiate the necessary resources it needs to satisfy the performance guarantees. Accordingly, the OS will grant (or not) to the run-time system the necessary resources it needs. Such a virtualization mechanism is very much in line with current integrated computing systems in embedded systems and server consolidation environments in the high-performance domain. In avionic systems, the integrated modular architecture (IMA) executes and manages several subsystems and each of these subsystems must run in isolation without being affected by the other subsystems, while the same philosophy is also followed in large clusters that are virtually decomposed into multiple independent systems.

1.6.2 Spatial and Temporal Component

Mapping virtual to actual resources is decomposed into a temporal and a spatial component. The spatial component specifies the fractions of the system's physical resources that are dedicated during the run time(s). For example, consider a baseline system containing four processors, each with a private L1 cache. The processors share an L2 cache, main memory, and supporting interconnection structures. Let us assume that the HARPA-RT distributes these resources among three applications with their corresponding run times (RT1, RT2, and RT3). After negotiating the SLAs, two cores are assigned to RT1, while RT1 and RT2 are assigned to a single core. The HARPA-OS engine assigns RT1 a significant fraction (50%) of the virtualized resources to support a demanding multithreaded application, and assigns the other two applications only 10% of the resources. These assignments leave 30% of the cache and memory resources unallocated. The HARPA-OS engine will distribute unallocated resources to improve overall resource utilization and optimize secondary performance requirements. The spatial component of the virtualized resources might contain multiple processors. These configurations are a natural extension of gang scheduling and support hierarchical resource management. For example, schedulers and resource management policies running within a run time can schedule and manage the run time's assigned resources as if the run time were a real multiprocessor machine. That is, the HARPA-OS engine can support recursive virtualization.

The temporal component is based on the well-established concept of ideal proportional sharing. It specifies the fraction of processor time (processor time slices) and the spatial component of the virtualized resources dedicates to each run-time system. As with spatial virtualized resources, the temporal component of virtualized resources naturally lends itself to recursive virtualization and hierarchical resources management, and excess temporal service might exist.

This spatiotemporal mapping of virtual to actual resources needs to accommodate the time-dependent changes to the system performance as a result of aging and variations. The HARPA-OS component is expected to reappropriate resources accordingly, in order to guarantee performance dependability throughout the system's lifetime and regardless of any failures that may occur as a result of aging prior to the nominal lifetime of the system. Once again, the marriage of proactive and reactive measures is key in ensuring performance dependability in either embedded, or HPC settings.

1.7 Interfaces with the Hardware: Monitors, Knobs, and Technology Models

Several interfaces have to be defined to provide appropriate monitors to (1) track application behavior against the target metrics and (2) define a feedback control algorithm that can predict the resources needed by the application to reach the performance-dependability requirements.

Note that one interesting aspect that will be studied in the book is how the low-level monitors relate to higher-level monitors. For example, a typical wireless application will require guarantees on the data throughput to ensure real-time processing. The data throughput is a coarse-grain performance requirement set by the application (or the user). Other peripheral requirements (directly affecting performance), such as power consumption, energy, or transient errors are typical coarse-grain requirements. However, monitors in the systems are more subtle and do not always directly map to the user requirements. Examples of such hardware monitors are cache misses, TLB misses, temperature sensors, cache coherence conflicts, and other monitors that are directly given by the system. In addition, we will have many software-related monitors that allow us to detect which application tasks are active, which are their main parameters, or which memory blocks or address ranges are activated. Relating such fine-grain monitors with the global behavior of the application and the system is a key research issue.

1.7.1   Monitors and Knobs

An important challenge facing HARPA is the choice of the right monitors and knobs and their correct implementation at the right level (circuit, architecture, OS, or language level), keeping the respecting overhead low. This is essential for an implementation in mainstream architectures. Therefore, the implemented monitors and knobs should be lightweight and should have no or negligible impact on the chip. The goal is to employ a cross-layer approach, whereby monitors and knobs throughout the system stack facilitate a comprehensive control strategy aimed at ensuring performance dependability throughout the system's lifetime. The objective in a hardware-level design is to develop a set of monitors that will allow the identification of the main sources of performance unpredictability in the system (e.g., power/energy envelope, wear-out, time-dependent variations, etc.) and the primarily affected hardware structures (core, caches, NoCs, etc.). Starting from the work in [15, 16], we developed a set of knobs that will allow the control of the execution of applications, hence providing dependable performance guarantees. The hardware overhead of monitors/knobs should be less than 10% in terms of hardware area and energy consumption, with respect to the chip. Respectively in software-level the target is to complement the set of the aforementioned monitors/knobs in the hardware level so that the complete hardware/software monitoring system can track the resources that lead to at least 90% of the unpredictability observed within the system. The hardware/software knob system will achieve the required performance guarantees (in both embedded systems and HPC systems) with a deviation of less than 10%. The performance overhead of the software monitors/knobs should be less than 10%, and this overhead should not jeopardize the performance guarantees required by the running applications, i.e., the additional latency should not be exposed to the user. Chapter 8 extensively discusses a set of novel monitors and knobs for energy-efficient and reliable NoCs [12, 14].

1.8 HARPA Applications

The applications that are exploited within the HARPA context address equally both the HPC and ES domains. Performance-dependability challenges are faced in both domains with different implications for each one. The methodologies that are created in the HARPA context (HARPA-OS, HARPA-RT, Knobs, and Monitors) have conformed to both HPC and ES contexts. The industrial partners of HARPA provided on-the-field applications and platforms, where the concepts developed.

1.8.1 Embedded Systems (ES) Domain

The first ES application is a spectrum sensing application that aims to explore the frequency spectrum to get information about free or unused frequency bands, in order, for instance, to perform radio-frequencies allocation. The application gathers samples for a large frequency band, performs filtering, and splits the signal in different frequency bands for further analysis. Then, for each frequency band, provision of statistical information is given, together with view of signals carriers and spectral footprint. Chapter 6 presents how HARPA-RT engine is exploited to ensure the performance dependability in such an application.

A second ES application brings "intelligence" to the field of landslide and rockwall monitoring by creating a smart-bridge for a network able to process data remotely, using machine learning and artificial intelligence techniques, collected using a wireless sensor network and on-site cameras. Landslides represent a major threat to human life, property and constructed facilities, infrastructure, and natural environment in most mountainous and hilly regions of the world. An innovative approach to the monitoring of landslides exploits multiple sensor sources like accelerometer data from distributed sensors and images from on-site camera to be processed in an on-site embedded device using machine learning techniques to create a specific model of the landslide in order to be able to understand the movements. Then, the system will send to a central server only the result enabling a faster response time, a more frequent analysis, and the deployment in area where network coverage is limited. The application behavior will change accordingly to the solar power and battery life available to the system in order to keep the system running with the core monitoring features to guarantee at least basic monitoring capabilities. This system is a good candidate to test the reliability and real-time capabilities of the HARPA OS with its deployment in a real-world test case. Detailed description of the application HARPA OS results can be found in Chap. 11.

1.8.2 High-Performance Computing Domain

In HPC domain, one application developed on an HPC platform is the FLOREON+ system (FLOods REcognition On the Net), which is a modular web-based system for environmental risk modeling and simulation in the Moravian-Silesian region of the Czech Republic. The application is used to simplify the process of disaster management and increase its operability and effectiveness. More details about the application can be also found in Chap. 11. The incorporation of the HARPA engine into the FLOREON+ disaster and flood management system nicely illustrates the synergistic behavior of the Operating System (OS) and how the proposed solution leverages monitors and knobs to provide performance dependability. It also demonstrates the reuse in the HPC domain of the OS engine concept, which is originally derived from the embedded systems domain. That way, the objectives of HARPA have been addressed through test cases that incorporate all technical effects of full system realization.

1.9 Conclusion

HARPA's dependable-performance objective enables performance, energy, lifetime, and cost improvements of platforms used both in embedded and HPC setups, including data centers. These improvements are direct ramifications of the mitigation and tolerance of the HARPA approach to time-dependent variations. With increasing prevalence of time-dependent variations, HARPA provides a cost-effective approach to ensure timing guarantees under various constraints for embedded platforms when running applications with mixed criticalities. The HARPA engine can be augmented to prioritize the resource allocation according to user-defined parameters per application. The HARPA approach is aiming to improve how to monitor and manage characteristics, which correspond to nonfunctional constraints, such as timing, power, and lifetime. Finally, HARPA aims to provide the abstraction that the hardware performance is invariant, i.e., it is dependable. This permits the application developers to remain agnostic of complex hardware peculiarities due to time-dependent variations.

References

1. Austin, T. (2006). Razor: A low-power pipeline based on circuit-level timing speculation. In *Proceedings of the 19th Annual Symposium on Integrated Circuits and Systems Design* (pp. 13–13). New York: ACM.
2. Bellasi, P., Massari, G., & Fornaciari, W. (2015). Effective runtime resource management using Linux control groups with the BarbequeRTRM framework. *ACM Transactions on Embedded Computing Systems, 14*(2), 39:1–39:17.

3. Borkar, S., Karnik, T., Narendra, S., Tschanz, J., Keshavarzi, A., & De, V. (2003). Parameter variations and impact on circuits and microarchitecture. In *Proceedings of the 40th Annual Design Automation Conference* (pp. 338–342). New York: ACM.
4. Bowman, K., Tschanz, J., Wilkerson, V., Lu, S.-L., Karnik, T., De, V., et al. (2009). Circuit techniques for dynamic variation tolerance. In *Proceedings of the 46th Annual Design Automation Conference* (pp. 4–7). New York: ACM.
5. Duranton, M., Black-Schaffer, D., Yehia, S., & De Bosschere, K. (2011). *Computing systems: Research challenges ahead: The HiPEAC vision 2011/2012.*
6. Gheorghita, S. V., Palkovic, M., Hamers, J., Vandecappelle, A., Mamagkakis, S., Basten, T., et al. (2009). System-scenario-based design of dynamic embedded systems. *ACM Transactions on Design Automation of Electronic Systems, 14*(1), 3.
7. Henkel, J., Bauer, L., Dutt, N., Gupta, P., Nassif, S., Shafique, M., et al. (2013). Reliable on-chip systems in the nano-era: Lessons learnt and future trends. In *Proceedings of the 50th Annual Design Automation Conference* (p. 99). New York: ACM.
8. Kulkarni, J. P., Kim, K., & Roy, K. (2007). A 160 mv, fully differential, robust schmitt trigger based sub-threshold SRAM. In *Proceedings of the 2007 International Symposium on Low Power Electronics and Design* (pp. 171–176). New York: ACM.
9. Le, H. Q., Starke, W. J., Fields, J. S., O'Connell, F. P., Nguyen, D. Q., Ronchetti, B. J., et al. (2007). IBM power6 microarchitecture. *IBM Journal of Research and Development, 51*(6), 639–662.
10. McNairy, C., & Bhatia, R. (2005). Montecito: A dual-core, dual-thread itanium processor. *IEEE Micro, 25*(2), 10–20.
11. Munaga, S., & Catthoor, F. (2012). Reliability-aware proactive energy management in hard real-time systems: A motivational case study. In *Technological Innovations in Adaptive and Dependable Systems: Advancing Models and Concepts* (pp. 215–225). Hershey: IGI Global.
12. Prodromou, A., Panteli, A., Nicopoulos, C., & Sazeides, Y. (2012). Nocalert: An on-line and real-time fault detection mechanism for network-on-chip architectures. In *Proceedings of the 2012 45th Annual IEEE/ACM International Symposium on Microarchitecture, MICRO-45* (pp. 60–71). Washington: IEEE Computer Society.
13. Simone Libutti, W., & Massari, G. (2016). Co-scheduling tasks on multi-core heterogeneous systems: An energy-aware perspective. *IET Computers and Digital Techniques, 10,* 77–84(7).
14. Zoni, D., Canidio, A., Fornaciari, W., Englezakis, P., Nicopoulos, C., & Sazeides, Y. (2017). Blackout. *Journal of Parallel and Distributed Computing, 104*(C):130–145.
15. Zoni, D., Colombo, L., & Fornaciari, W. (2018). Darkcache: Energy-performance optimization of tiled multi-cores by adaptively power-gating LLC banks. *ACM Transactions on Architecture Code Optimization, 15*(2), 21:1–21:26.
16. Zoni, D., Flich, J., & Fornaciari, W. (2016). Cutbuf: Buffer management and router design for traffic mixing in VNET-based NoCs. *IEEE Transactions on Parallel and Distributed Systems, 27*(6), 1603–1616.

Part I

Chapter 2
Trends in Processor Architecture

Antonio González

2.1 Past Trends

Processors have undergone a tremendous evolution throughout their history. A key milestone in this evolution was the introduction of the microprocessor, term that refers to a processor that is implemented in a single chip. The first microprocessor was introduced by Intel under the name of Intel 4004 in 1971. It contained about 2300 transistors, was clocked at 740 kHz, and delivered 92,000 instructions per second while dissipating around 0.5 W.

Since then, practically every year we have witnessed the launch of a new microprocessor, delivering significant performance improvements over previous ones. Some studies have estimated this growth to be exponential, in the order of about 50% per year, which results in a cumulative growth of over three orders of magnitude in a time span of two decades [12]. These improvements have been fueled by advances in the manufacturing process and innovations in processor architecture. According to several studies [4, 6], both aspects contributed in a similar amount to the global gains.

The manufacturing process technology has tried to follow the scaling recipe laid down by Robert N. Dennard in the early 1970s [7]. The basics of this technology scaling consists of reducing transistor dimensions by a factor of 30% every generation (typically 2 years) while keeping electric fields constant. The 30% scaling in the dimensions results in doubling the transistor density (doubling transistor density every 2 years was predicted in 1975 by Gordon Moore and is normally referred to as Moore's Law [21, 22]). To keep the electric field constant, supply voltage should also be reduced by 30%. All together would result in a 30%

A. González (✉)
Universitat Politècnica de Catalunya, Barcelona, Spain
e-mail: antonio@ac.upc.edu

© Springer International Publishing AG, part of Springer Nature 2019
W. Fornaciari, D. Soudris (eds.), *Harnessing Performance Variability in Embedded and High-performance Many/Multi-core Platforms*,
https://doi.org/10.1007/978-3-319-91962-1_2

reduction in delay and no variation in power density. If total area of the chip is kept constant, the net result is twice the number of transistors that are 43% faster, and the same total power dissipation.

More transistors can be used to increase the processor throughput. Theoretically, doubling the number of transistors in a chip provides it with the capability of performing twice the number of functions in the same time, and increasing its storage by a factor of two. In practice, however, performance gains are significantly lower. Fred Pollack made the observation long time ago that processor performance was approximately proportional to the square root of its area, which is normally referred to as Pollack's rule of thumb [24]. This is mainly due to the following two reasons. First, the internal microarchitecture of processors: the performance of many of its key components such as the issue logic and cache memories does not scale linearly with area. Second, even if transistors are smaller by a factor of two, this has not resulted in twice the number of transistors per unit of area due to the increasing impact of wires. Having more functional blocks often requires an increase in the number of wires that is superlinear, which increases the percentage of area that must be devoted to them. An example of this is the bypass logic. Having a full bypass among all functional units requires the number of wires that grows quadratically with the number of units.

On the other hand, the 30% reduction in transistor delay has the potential to make circuits 43% faster. However, benefits in practice are lower, due to the fact that wire delays do not scale at the same pace [13], and because of that, they have become the main bottleneck for microprocessor performance. Besides, as processor structures become more complex, a larger percentage of the area must be devoted to wires, as outlined above, so the impact of wire delays become even more severe. As a result, the time spent in moving data around is the main component of the activity performed by current microprocessors.

Putting it all together, the increased transistor density allows architects to include more compute and storage units and/or more complex units in the microprocessors, which used to provide an increase in performance of about 40% per process generation (i.e., Pollack's rule of thumb). The 30% reduction in delay can provide an additional improvement, as high as 40% but normally is lower due to the impact of wire delays. Finally, additional performance improvements come from microarchitecture innovation. These innovations include deeper pipelines, more effective cache memory organizations, more accurate branch predictors, new instruction set architecture (ISA) features, out-of-order execution, larger instruction windows, and multicore architectures just to name some of the most relevant. This resulted in a total performance improvement rate of 52% per year during the period 1986–2003 [12] as measured by SPEC benchmarks. However, this growth rate has dropped to about 22% since 2003 [12].

Multiple reasons have contributed to the slowdown in performance improvement. First, supply voltage has not scaled as dictated by Dennard's recipe; in fact, its decrease has been much lower, in the order of 8% per year (15% every process generation, assuming a 2-year process technology cadence), as shown in Fig. 2.1.

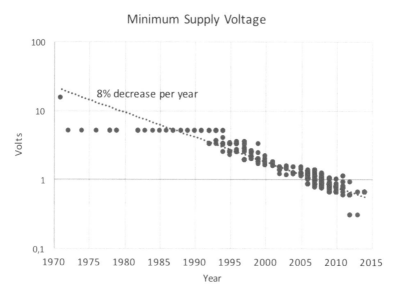

Fig. 2.1 Minimum supply voltage of microprocessors

The main direct consequence of scaling supply voltage by less than dictated by Dennard's guidelines is an increase in power density and total power of microprocessors. As can be seen in Fig. 2.2, during 1980s and 1990s microprocessor power increased in an exponential manner, by about two orders of magnitude in two decades. An obvious consequence of power increase is an increase in energy consumption and thus in the operating cost of computing systems. More importantly, this also implied a similar increase in power density since microprocessor area has not changed much over the years. As pointed out by some authors [3], microprocessors in 0.6-μm technology (in the early 1990s) surpassed the power density of a kitchen hot plate's heating coil, and the trend continued increasing by about ten more years, reaching unsustainable levels. Increased power density requires a more powerful cooling solution since power density is a proxy of heat dissipation. Temperature has an important impact on reliability and leakage currents so silicon operating temperature must be kept below a certain limit (in the order of 100 °C), which may be unaffordable in some systems due to either its cost or physical characteristics (volume, weight, noise, etc.).

Microprocessor's power growth resulted unsustainable in the early 2000s, and the trend changed towards flat or decreasing power budgets. This explains in part the inflexion point in the performance growth curve which coincides in time with this change in power budget. To keep power density constant when supply voltage scales less than transistors dimensions, the only way is to have a lower percentage of transistors switching at any given time. This motivated the aggressive use of power-aware techniques such as clock gating [34] and power gating [15], which are extensively used by today's microprocessors. Despite the great benefits of these

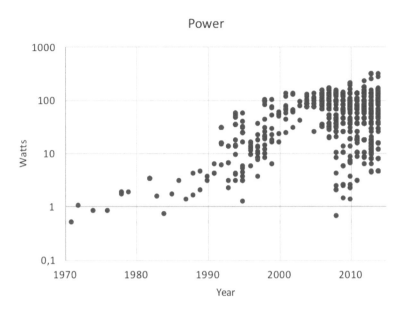

Fig. 2.2 Power (TDP) of microprocessors

techniques, they have not been able to provide the exponential benefits that would be needed to compensate for the difference between desired and actual supply voltage scaling (30% vs 15%). This has led architects to include techniques to avoid all blocks to be active at the same time, which sometimes is referred to as "dark silicon" [8], or to avoid all blocks to operate at maximum supply voltage through extensive use of dynamic voltage scaling techniques. For instance, in some contemporary processors with multiple cores, when all cores are active supply voltage cannot reach the same level as when only one of the cores is active.

On the other hand, the second important reason that explains the performance inflexion point observed in the early 2000s is due to a slowdown in improvements coming from architecture. During the 1980s and 1990s, there were many innovations in processor architecture, including deep pipelining, out-of-order execution, branch prediction, data prefetching, superscalar, and memory hierarchy improvements, among others. All these techniques provided significant improvements in performance by either reducing the latency of main functions (especially memory operations) or increasing the degree of instruction-level parallelism (ILP). After two decades of continuous improvements in these areas, they reached a point of diminishing returns. This prompted the inclusion of techniques to exploit thread-level parallelism (TLP) at the processor level, and gave birth to multicore and multithreaded processors, which are extensively used nowadays. TLP techniques have proved to be very effective to increase processor throughput when the workload consists of independent applications. However, they are often less effective when it comes to decompose a single application into a number of parallel threads. Serial parts of the application limit the benefits of TLP, as stated in 1967 by Gene Amdahl [1]. Besides, other important aspects such as workload balance, synchronization, and communication cost also pose significant hurdles to the benefits of TLP.

2.2 Current Microprocessors

Figure 2.3 shows a high-level block diagram of a typical contemporary microprocessor. The main components are a number of general-purpose cores (4 in the figure), a graphics processing unit (GPU), a shared last-level cache, a memory, and I/O interface, and an on-chip fabric to interconnect all these components. Below, we briefly describe the architecture of these modules.

2.2.1 General-Purpose Cores

The architecture of general-purpose cores has significantly evolved throughout its history. From the very early processors, pipelining and cache memories have been the two key microarchitectural techniques used to improve performance.

ILLIAC II [5] and IBM Stretch [2] computers were pioneers in the use of pipelining. Pipelining is an extremely cost-effective technique to exploit ILP, and has been used by practically all microprocessors. A pipeline with N stages can potentially overlap the execution of N instructions and, thus, provide an increase in throughput by a factor of N. In practice, benefits are lower due mainly to instruction dependences, which introduce bubbles in the pipeline. Initial processors had just a few pipeline stages but the introduction of RISC architectures in the early 1980s made pipelining to be more cost-effective and facilitated the use of deeper

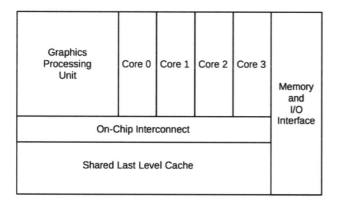

Fig. 2.3 High-level block diagram of a contemporary microprocessor

pipelines. Innovations in microarchitecture techniques to avoid pipeline bubbles due to dependences, such as out-of-order execution, data dependence speculation, and branch prediction, allowed a continuous increase in pipeline depth, during the 1980s and 1990s. Some studies in the early 2000s suggested that the optimal pipeline depth for performance could be in the order of 50 stages [14, 28], but some later studies [10, 29] showed that when power is taken into account, the optimal pipeline depth is much shallower, and in practice, pipeline depth increase stopped at around 20 stages. Energy efficiency favors shallow pipelines since deeper pipelines consume more energy due to extra latching, more complex control, and energy waste due to speculative execution techniques. As a result, current microprocessor pipelines normally have between 10 and 20 stages and the forecast for the future is that pipeline depths will keep around this range.

Microprocessors require many accesses to memory to execute a program. Since instructions and data are stored in memory, a typical instruction requires four memory accesses, one to fetch the instruction, two to read its two source operands, and one to write its result. While processor performance has increased at a rate ranging from 22% to 52% per year, as commented above, memory latency has improved at a much lower rate of about 7% per year. In a period of 40 years, this resulted in a gap of four orders of magnitude [12] . This trend motivated the organization of memory in a hierarchical way. A hierarchical organization exploits a core feature of the technology: smaller blocks are normally faster. This applies to all kind of logic blocks, and in particular to memories: a small memory is faster than a larger one. This is basically due to the fact that latency of read and write operations is dominated by wire delays of the word and bit lines. The smaller the memory, the shorter the word lines and bit lines are.

A hierarchical organization of memory was used as early as the mainframes of the 1960s, which used a hierarchy of physical memory composed of modules built with different technologies: semiconductor, magnetic core, drum, and disk. Data caches were used for the first time in the IBM 360 Model 85 computers [16]. Most computers nowadays have a memory hierarchy organized in six levels: a register file, three levels of cache memory, main memory, and disk.

Another important microarchitecture technique used since the very early processors has been branch prediction. When a branch is fetched, the next instructions to be executed depend on the branch outcome, which is not available until the branch is executed some cycles later. Stalling the fetch until the outcome of the branch is known would cause a significant penalty in pipelined processors since branches are very common in many programs (in the order of one out of ten instructions for nonnumerical codes). Predicting the outcome of the branch and speculatively executing the instructions of the predicted path can alleviate this penalty provided that the predictor is highly accurate. This technique is known as branch prediction and was used as early as the late 1950s in the IBM Stretch computer [2].

Branch predictors based on a table of two-bit saturating counters were introduced by James Smith in the late 1970s [26] and have been used in practically all microprocessors since then. Since the penalty of a branch misprediction grows linearly with the pipeline depth, deeper pipelines prompted research on more sophisticated predictors such as the two-level branch predictor proposed by Yeh and Patt in the early 1990s [35], variants of which have been used by many modern microprocessors. These predictors are based on two-bit saturating counters, but the particular counter used in each case depends not only on the particular branch being predicted but also on the outcome of recent past branches. Another important innovation in the area of branch prediction that has been used by many microprocessors is the concept of hybrid predictors, first introduced by Scott McFarling in 1993 [20]. The key idea is based on two observations: first, the most cost-effective branch predictor is different for different types of codes; second, more sophisticated predictors tend to have a longer warm-up time, so at the beginning of each context-switch interval simple predictors tend to be more accurate than complex ones, whereas this trend is inverted once the complex predictor has been warmed-up. Based on these observations, he proposed a mechanism to combine multiple predictors and dynamically choose which one is more likely to be the most accurate.

The ISA is another important feature of a processor architecture. The ISA has evolved over the years to include new instructions that can better support common code constructs in mainstream application domains. An important change in the design of ISA was introduced in the early 1980s, mainly driven by independent projects at UC Berkeley [23], IBM [25], and Stanford University [11]. This trend was coined as RISC (reduced instruction set computer) by the team in UC Berkeley, term that was later adopted by everyone else. Up until then, the trend in ISA design was to increase its complexity. It was believed that by closing the semantic gap between high-level languages and machine languages, computer could be made to be more efficient. These projects demonstrated that a much simpler ISA could be competitive, or even outperform, a more complex one, and at the same time offer additional advantages such as a much lower cost of design and verification of the processor. The maturity of compiler technology combined with the benefits of circuit simplicity in terms of delay, cost, and energy consumption were some of the key aspects that favored the success of RISC ISAs.

Enhancements in pipelining, amplified by the simplicity introduced by RISC ISAs, caching, and branch prediction were the main driving forces behind microarchitecture innovation during the 1980s. These processors aimed to achieve a throughput of one instruction per cycle, which they approached after several generations of improvements. During the 1990s, a new microarchitecture organization to exploit further ILP became common use, which was coined superscalar processors. A superscalar processor is a processor that can process multiple instructions in all its pipeline stages. That is, it can fetch, decode, rename, issue, execute, and commit multiple instructions at the same time. How many instructions can be processed in parallel in each pipeline stage is referred to as the width of the superscalar processor. In practice, not all pipeline stages may have the width, so we define the width of the processor as the minimum width of its pipeline stages. An N-wide superscalar processor can potentially achieve a performance of N instructions per cycle.

A superscalar processor requires to replicate hardware resources, normally by the same factor as its width, but in some parts of the processor, the hardware cost may grow superlinearly with the width. An example of this superlinear cost is the bypass logic, which grows quadratically with the number of functional units. Another not so obvious example is the issue logic. In this case, the reason is that to find more independent instructions to be issued every cycle, the hardware has to search in a larger instruction window, so if we double the issue width, we need twice the number of issue ports, but each port has to search in a much larger window. Superscalar width grew during the 1990s, but due to this superlinear cost, and the difficulties to find enough ILP, it flattened out to around 4, and practically all current processors have a superscalar width between 2 and 8.

Another key microarchitecture technique that became common use during the 1990s was out-of-order execution, also known as dynamic scheduling. The main idea is to allow the hardware to execute the instructions in an order different to the one that they appear in the binary, with the constraint that the semantics of the program must not be changed. The goal is to find more ILP by reordering the instructions. For instance, the consumer of a load that misses in cache may be stalled during many cycles. Instead of stalling all instructions younger than this consumer, as in-order processors do, out-of-order processors can execute younger instructions, provided that they do not depend on the load. Out-of-order execution provides important benefits in performance but has also an important cost, mainly due to a much more complex issue logic, including the issue logic of memory instructions, which normally is separated from the issue logic of the rest of instructions, and is quite costly since memory dependences are more difficult to check than register dependences. This is due to the fact that register dependences are known at decode time and can easily be identified as instructions are decoded/renamed in program order whereas memory dependences need to be identified later in the back end of the pipeline, once the effective addresses of loads and stores are computed, and this computation is performed out of program order.

Out-of-order execution became popular in the 1990s and is used by the vast majority of current microprocessors. However, the main ideas behind this technique date back to the 1960s. In fact, out-of-order execution was first introduced by the CDC 6600 computer [31] in 1964 and was later improved by the IBM System/360 Model 91 [32]. IBM approach is usually referred to as Tomasulo's algorithm, and is the bases for the schemes used nowadays. The main innovation introduced by IBM over the CDC scheme was the use of register renaming. By renaming register operands, the processor has more options to reorder the code, since it only needs to respect data dependences (a.k.a. read-after-write dependences). Name dependences (a.k.a. write-after-write and write-after-read dependences) are removed and do not impose any ordering constraint. Register renaming provides huge benefits in terms of ILP and, thus, is used by all current out-of-order processors to the best of our knowledge.

During the 2000s, the main architectural innovation introduced at the micropro-cessor core level is called multithreading. There are different variants of the concept of multithreading, but the most commonly adopted by current microprocessors is known as simultaneous multithreading [33] and it was first used by commercial processors in the Intel's Pentium 4, under the name of hyperthreading [17]. The key idea behind this technique is to provide a microprocessor core with the capabilities to execute multiple threads simultaneously sharing the majority of hardware resources. At any given cycle and any given pipeline stage, the processor can potentially execute multiple instructions belonging to different threads. In this way, the processor can remove some pipeline bubbles and the multiple slots of a superscalar processor in each pipeline stage can be more frequently filled with useful work. In other words, multithreading increases the utilization of the resources already present in a pipelined, superscalar processor.

This technique is highly efficient from the hardware cost point of view, since it requires small extensions to a conventional superscalar processor. Its main cost is an increase in register file storage, since the processor has to keep the state of multiple threads simultaneously. On the other hand, for this same reason, its scalability is limited. Since practically all processor core resources are shared among the simultaneously running threads, there are frequent structural hazards that prevent each thread to run at the same speed as it would if it run alone.

The benefits of increasing the degree of multithreading diminish quickly for general-purpose cores. With two to four threads, resources get highly utilized, and adding more threads would bring minimal benefits in many cases. Because of that, the typical multithreading degree of general-purpose cores is in the range of two to four, although for server processor it may be a bit higher, such as the recently announced IBM Power9 [30] that supports up to eight simultaneous threads. Server workloads are sometimes highly memory intensive, and the utilization of the CPU resources is low, which allows the possibility to exploit a larger number of simultaneous threads.

Another type of processing unit that is highly multithreaded is the GPU. In this case, GPUs are designed to support a very large number of threads since graphics applications exhibit a huge degree of TLP (e.g., most of the operations needed to compute the color of each pixel are performed by independent threads). GPUs rely on a high degree of multithreading to hide memory latencies rather than on a complex memory hierarchy. Even if threads are frequently stalled due to long latency memory accesses, there are normally other threads ready that can use the hardware resources to make progress and keep the hardware highly utilized.

To conclude this section, Fig. 2.4 shows a high-level overview of the typical microarchitecture of modern general-purpose cores.

Instructions are fetched from an on-chip instruction cache, which normally has a few tens of kilobytes. The main component of the instruction fetch logic is a branch predictor. Multiple instructions can be fetched in the same cycle by using a single, wide cache memory port that can provide multiple consecutive bytes in a single access. Branch prediction is performed in parallel with the instruction cache access (or even before in some processors) to avoid bubbles in the pipeline (otherwise, in the next cycle the processor would not know which are the next set of instructions to fetch). Multiple fetched instructions can potentially be branches, so the branch predictor needs to predict multiple branches in parallel. However, depending on the branch predictor being used this may not be that easy. In particular, if the branch predictor uses global history, when predicting a branch the outcome of the

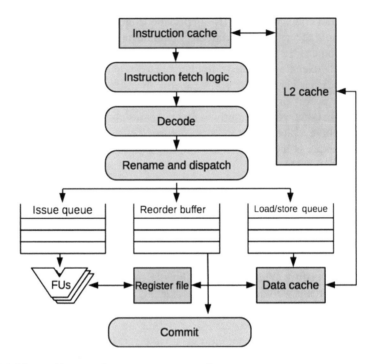

Fig. 2.4 Microarchitecture of a contemporary general-purpose core

previous branches must be known. This is typically solved by assuming that the previous branches being predicted in parallel are not taken. In case any of the branch predictions turns out to be taken, all the following branch predictions are discarded. Since all instructions fetched in parallel are consecutive, all instructions after the predicted taken branch are discarded too, since they are not in the predicted path. Thus, the branch predictor can normally predict multiple branches per cycle but only up to one taken branch.

After being fetched, instructions are processed by the decode logic, which identifies their type, which operands are required, and other relevant fields of the instruction. Decoding multiple instructions in parallel does not pose any challenge and simply requires to replicate hardware since each instruction can be decoded independently of the others. Some processors perform a dynamic translation at this stage. This is common in processors using old ISAs such as x86, in order to translate the compiler-generated instructions to an internal ISA which is more amenable for pipelining and other microarchitectural optimizations. This translation is relatively simple since it is done for each instruction individually; optimizing blocks of instructions would be more effective but also much more complex and is normally not used.

Afterwards, instructions rename their register operands. As outlined above, this step is used by out-of-order processors to remove name dependences and increase in this way the amount of ILP. In-order processors normally do not make use of it. Renaming is done basically through a table that keeps track of the latest storage location (a.k.a. physical register) assigned to each logical register. Source operands are translated to physical registers by reading the corresponding entries in the renaming table. The destination register of each given instruction is assigned to a free physical register (if there are no physical registers available, renaming is stalled). Superscalar processors rename multiple instructions per cycle. This requires a rename table with multiple read/write ports, and some additional logic to take into account the data dependences among instructions being renamed in parallel. In particular, each source operand must be checked with all previous destination operands, and in case of a match, the physical register identifier used by this source operand is not the one coming from the renaming table but the one assigned to the closest matching destination operand. Next, instructions are dispatched to the issue queue and the reorder buffer, and memory instructions are also dispatched to the load/store queue.

All the above operations are performed in program order in all processors, including out-of-order processors. This part of the pipeline is normally referred to as the front end. An in-order front end simplifies the detection of register data dependences among instructions, but on the other hand, exacerbates the penalty of any potential stall. Most harmful stalls in the front-end pipeline are due to instruction cache misses.

Instructions remain in the issue queue until they are selected to be executed. The process to select the instructions to be executed every cycle is called instruction issue. For in-order processors, this logic is relatively simple. To issue N instructions per cycle, it just needs to check the N oldest instructions in the issue queue. Each of these instructions is issued if its source operands are ready and the required resources for its execution are available (e.g., functional units, register file ports, etc.).

Most processors nowadays can issue instructions out-of-order. This is much more complex since it requires to check all instructions in the issue queue, since all of them are candidates to be issued. To this end, the issue queue stores the renamed source operands of each instruction and a ready bit for each of them. Every time an instruction is executed, its destination register id is checked against all source operands in all entries, and for every match, the corresponding ready bit is set (this action is normally referred to as wake-up). Instructions that have all source operands ready, and the required resources are available are selected to be executed at every cycle. If there are more candidate instructions than available resources, a heuristic is applied to prioritize the candidate instructions. Given priority to the oldest instructions is quite common, and sometimes this is combined with other simple heuristics such as prioritizing long latency instructions.

Each issued instruction reads first its source operands, then is executed in a functional unit, and finally writes its result in the register file. The register file is therefore highly multi-ported, to support multiple reads and writes per cycle. For instance, a four-way issue processor may typically have eight read ports and four write ports. Processors have multiple functional units to execute multiple operations in parallel, including integer ALUs and floating-point ALUs. Besides, practically all processors include a rich set of SIMD (single-instruction, multiple-data) instructions in their ISA. These are instructions that operate on vector operands stored in registers and are usually known as multimedia extensions, since they are very effective for multimedia applications. To support these instructions, processors also include SIMD units, which can process all the elements of a vector operation in parallel. Since there are three different types of register operands, i.e., integer, floating-point, and vector, some processors have three separate register files, one for each type, whereas others have only two, one for integer and another for floating-point and vector.

Instructions wait in the reorder buffer until they commit. The reorder buffer stores some bookkeeping data for each in-flight instruction plus the information required in case the instruction needs to be squashed. An in-flight instruction may be squashed for a variety of reasons such as an older branch being mispredicted, or an older instruction generating an exception. Instructions commit in program order, even if the processor is out-of-order, since an easy way to guarantee that an instruction will not need to be squashed is waiting for the commit of all older instructions. At this point, no older instruction may cause an exception, branch misprediction, nor any other type of event.

Memory instructions deserve special attention. To be issued, they must go through a similar process as other instructions. However, the actions to be performed are a bit more complex since dependences are more difficult to identify because they involve memory locations rather than registers. Register operands' ids, and therefore register dependences, are known during renaming, but memory addresses, and thus memory dependences, are not. The address of the memory location read or written by a memory instruction is normally computed at run time by adding an offset to the content of a register. Until this register operand is not ready, the address cannot be known. The logic associated with the load/store queue is responsible for computing these addresses, checking for potential dependences and deciding when memory reads can be issued.

Store instructions do not write into memory until they commit. This is because a write to memory cannot be undone, unlike writes to registers, since they may be visible to other processes. Once a write is visible to another process, it may trigger some activity in this process (e.g., entering a critical region), which could not be undone if the store turned out to be squashed.

A load instruction can safely read from memory as soon as its effective address has been computed and there are no dependences with older stores. Checking for dependences requires to compare memory addresses, which are much longer than register ids, and besides the check has to take into account the potential different data widths of loads and stores.

To guarantee that a load does not depend on any older store, it must wait until the effective addresses of all older stores are computed. This may significantly delay a load, and most of the times this delay is unnecessary since memory dependences with instructions close enough to be in the instruction window at the same time are not common. To avoid this delay, some processors use some kind of mechanism to predict memory dependences and just stall loads that are likely to have a dependence. This family of mechanisms is called data dependence speculation [9, 19], and it requires a recovery scheme for mispredicted loads, similar to the one used for branch mispredictions.

Store-to-load forwarding is commonly used when a memory dependence is encountered/predicted between a load and a not committed store. Rather than waiting for the store to write to memory, the load gets the data directly from the store.

2.2.2 Multicore Processors

The vast majority of current microprocessors have multiple general-purpose cores based on the architecture described in the previous section. Figure 2.5 depicts the main components of a multicore processor. There are a number of cores, each one with private L1 caches (separate caches for instructions and data) and a private second-level cache (some processors do not have a private L2 cache), a shared last-

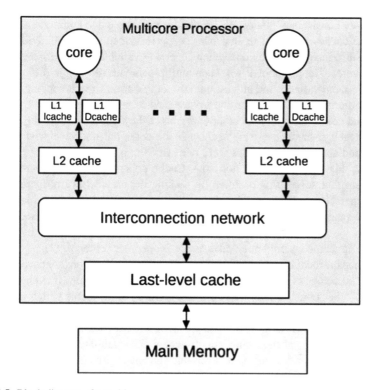

Fig. 2.5 Block diagram of a multicore processor

level cache, and an interconnection network that allows all the cores to communicate through the memory hierarchy. The lower levels of the memory hierarchy are normally a main memory and a disk storage, and they are located off-chip. A multicore processor can run multiple threads simultaneously in different cores with no resource contention among them except for the shared resources, which are basically the memory hierarchy and the interconnection network. The architecture of these two components is key for the performance of multicore processors and is described in more detail below. The architecture of each one of the individual cores is basically the same as in a single-core processor, and has been described in the previous section.

2.2.2.1 Memory Hierarchy

Parallel threads running in different cores can share the memory address space, which is a very desirable feature for programmability. However, this introduces two important challenges: how to keep data coherent when it is used by multiple threads in different cores, and how to make memory operations visible to the other threads running simultaneously in other cores. The former challenge is solved by a new mechanism that is called cache coherence. The latter is handled by what is called the memory consistency model. We briefly describe these two components below.

Private caches are great to improve the performance of each individual core, but they introduce some problems when accessing variables that are shared among multiple cores. The problem arises when multiple threads running in different cores access the same variable and at least one of them performs a write operation. Since variables are normally copied in the private caches, a write performed by a given core would not be visible to the other cores, since they would keep seeing the value stored in their respective private caches. If the cores did not have private caches and all read and write operations were performed in a completely shared memory hierarchy, this problem would not exist. Cache coherence is a microarchitecture mechanism that solves this problem by making private caches transparent to the programmers. In other words, with cache coherence, a processor with private caches produces exactly the same results as the same processor with all private caches removed.

Cache coherence is implemented through a coherence protocol. The basic idea of a coherence protocol is to make sure that writes from a given core are propagated to all potential copies of the same variable in the memory hierarchy, including private caches of other cores. There are two main families of protocols which are called write-invalidate and write-update protocols, respectively, the former being the more common solution. In a write-invalidate protocol, when a core has to perform a write operation, all copies of the same data elsewhere are invalidated before performing the write. If other cores are using this variable and have a private copy, they will have to request it again after being invalidated, and will get the updated value from the core that performed the write. In write-update protocols, a write operation updates all copies of the variable in the system. They are more complex than write-invalidate protocols, and consume more network bandwidth, thus they are rarely used.

The implementation of a cache coherence protocol is complex, and requires a careful design and a thorough validation to guarantee that it is correct. The complexity comes from the fact that a transaction (e.g., an invalidation request) generated by one core has to be propagated to the other cores and this cannot be done instantaneously. In the meantime, while a transaction is being processed, other request to the same variable may occur in the system, and the system has to deal with them and guarantee correct semantics and no deadlocks. This is normally solved by having a number of stable states that represent the state of a memory block after a transaction has been completed plus a number of transient states that represent the state in different phases of an ongoing transaction. The memory block granularity normally used to keep state information is a cache memory block. The most typical stable states used are the following:

- Invalid: the block is not present or is stale.
- Shared: The block can be read but not modified.
- Exclusive: The block can be read, and it is the only valid copy.
- Modified: The block can be read and written.

There are two main families of protocols regarding how the state information is stored and how transactions are processed. In a snoopy-based protocol, a transaction generated by a cache controller is broadcast to all other controllers. The system relies on the network to guarantee that all messages arrive to all cores in a consistent order. For instance, most protocols assume that they arrive in the same order to all cores, and use a shared bus to guarantee this order. The other family of protocols is called directory-based. In this case, there is a directory that holds the state of each memory block and normally sits next to the LLC. A transaction generated by a cache controller is sent to the directory, and based on the information in the directory, the request is served by the LLC or it is forwarded to the corresponding private cache(s). In general, snoopy-based protocols are simpler but directory-based protocols are more scalable and, thus, more effective for a large number of cores.

The memory consistency model is a formal definition of which potential executions (i.e., outcomes) of a parallel program are allowed and which ones are not. Note that multiple different outcomes for a given program may be allowed since many parallel programs are nondeterministic. In other words, the memory consistency model specifies what values load instructions may return and what is the final state of the memory for a given input data. This information is needed by the programmers in order to write programs that do what they expect, so it is part of the ISA.

For instance, assume a code in which two threads modify a different variable each, and later they read the variable modified by the other thread. If there are no synchronization operations in between, multiple outcomes are possible. The thread that executes ahead of the other will get the old value, whereas the other thread will get the updated one. Alternatively, both threads can end up reading the updated value if the two threads perform the two writes before the two reads. All systems allow these three alternative outcomes but some systems also allow a less intuitive one, which is that the two threads read the old value. The memory consistency model precisely defines which of these four outcomes is allowed, so that the programmer knows what to expect when the code is executed.

Different memory consistency models represent a different trade-off between programmability and performance. The most intuitive consistency model for programmers is called sequential consistency and was introduced by Lamport in 1979 [18]. A system is sequentially consistent if "the result of any execution is the same as if the operations of all processors (cores) were executed in some sequential order, and the operations of each individual processor (core) appear in this sequence in the order specified by its program." This memory model is the most intuitive for programmers but is also the one that imposes more constraints to the hardware in terms of when memory accesses can be performed.

Another widely used memory model nowadays is called total store ordering, which is used by SPARC and x86 processors [27]. For most program idioms, it behaves like sequential consistency, but it allows some new executions not permitted by sequential consistency. For instance, for the code example above, total store ordering allows both threads to read the old value, whereas sequential consistency does not. More precisely, total store ordering imposes the same ordering constraints as sequential consistency with one exception: loads can be moved above older non-conflicting (to different addresses) stores. The motivation for removing this constraint is performance. By removing it, the processor can use a write buffer to hold committed stores until they get read–write permissions. This buffer is very effective to hide the memory latency for store misses, so it results in significant performance improvements and is widely used by microprocessors.

2.2.2.2 Interconnection Network

As illustrated in Fig. 2.5, a multicore processor consists of a number of cores and a last-level cache that need to communicate among them. This communication is carried out through an interconnection network. In general, each core and its private caches are a node in this network, and the last-level cache is another node of the network. It is also common that the last-level cache is split into multiple modules, each one being a different node of the network. Each one of these nodes is connected to a switch through a network interface. Each switch has a number of links that connect it to other switches as shown in Fig. 2.6.

The different switches are interconnected following a particular topology. For instance, Fig. 2.7 shows a mesh topology, which is one of the most commonly used topologies for multicore processors. A line is another common topology, which is basically a mesh with just one node in one of the two dimensions. If the two nodes at the end of the line are interconnected, then we have a ring for the one-dimensional case or a torus for a two-dimensional case, which are also commonly used.

Fig. 2.6 Block diagram of network node

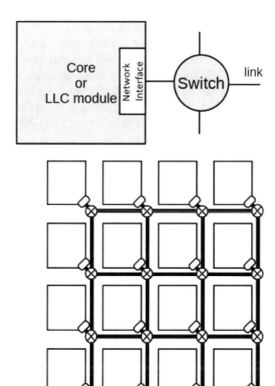

Fig. 2.7 A mesh network topology

2.3   Going Forward: Specialized Units

We are witnessing a slowdown in process technology improvements. Supply voltage has not scaled as much as Dennard's guidelines [7] for years and Moore's Law scaling [21, 22] has slowed down in recent years and is approaching a limit unless some breakthrough in process technology appears in the coming years. Chip manufacturing technology has been based on silicon for decades. Current process technology uses a 10-nm fabrication process and the minimum feature size of circuit components is approaching atomic dimensions. Taking into account that silicon lattice spacing is around 0.5 nm, the current 10 nm feature size is equivalent to around 20 atoms. Further miniaturization is going to be very difficult to achieve and dimension scaling will very likely stop in a few more generations.

On the other hand, increasing energy efficiency will keep being a main driving force for innovation in computing systems. This is easily explained by observing that the power dissipation of a system is equal to the average energy consumed per task multiplied by the number of tasks that the system can perform by unit of time. Note that the latter term is the performance of the system whereas the former is what we call energy efficiency. If power cannot be increased due to the various reasons explained in Sect. 2.1 of this chapter, any increase in performance requires a reduction in energy per task of the same magnitude. In other words, the challenge for future computing systems is not only how to make them run faster, but at the same time, this increase in performance has to come with a similar decrease in energy consumption (i.e., a similar improvement in energy efficiency). This is really challenging, since normally higher throughput implies higher energy. For instance, a faster addition requires to increase the supply voltage of the adder or to use a more sophisticated adder. In both cases, the energy to perform an addition increases.

Since process technology is unlikely to provide significant additional benefits in the future, the improvements in energy efficiency have to come from other areas such as microarchitecture. However, the microarchitecture of general-purpose cores has been optimized for several decades, and there is little headroom for further improvements. Perhaps the most promising approach is "specialization." It is well known that a specialized unit can provide dramatic improvements in energy efficiency when compared to a general-purpose unit. Specialized units are already a reality in most processors. For instance, practically all current processors include a GPU, which is a unit specialized on image rendering. Most processors for smart phones include specialized units for image and audio processing.

In the future, we are likely to see an explosion in the use of specialized units. The drawback of specialization is its higher cost in comparison with general-purpose units, since their cost needs to be amortized over a smaller number applications. However, the flip side of the death of Moore's Law will be a significant decrease in cost of chip fabrication, since manufacturing industries will not need to replace their equipment so often, and their investments in process technology will be drastically reduced. Under this scenario, transistors will be very chip, much chipper than they already are, and this will open the door for many new opportunities to use them in specialized units.

References

1. Amdahl, G. M. (1967). Validity of the single processor approach to achieving large scale computing capabilities. In *Proceedings of the Spring Joint Computer Conference AFIPS* (pp. 483–485).
2. Bloch, E. (1959). The engineering design of the stretch computer. In *Proceedings of the Eastern Joint IRE-AIEE-ACM Computer Conference* (pp. 48–58).
3. Borkar, S. (1999). Design challenges of technology scaling. *IEEE Micro, 19*(4), 23–29.
4. Borkar, S., & Chien, A. A. (2011). The future of microprocessors. *Communications of the ACM, 54*(5), 67–77.
5. Brearley, H. (1965). ILLIAC II-A short description and annotated bibliography. *IEEE Transactions on Electronic Computers, EC-14*(3), 399–403.
6. Danowitz, A., Kelley, K., Mao, J., Stevenson, J. P., & Horowitz, M. (2012). CPU DB: Recording microprocessor history. *Communications of the ACM, 55*(4):55–63.
7. Dennard, R. H., Gaensslen, F. H., Rideout, V. L., Bassous, E., & LeBlanc, A. R. (1974). Design of ion-implanted MOSFET's with very small physical dimensions. *IEEE Journal of Solid-State Circuits, 9*(5):256–268.

8. Esmaeilzadeh, H., Blem, E., Amant, R. S., Sankaralingam, K., & Burger, D. (2011). Dark silicon and the end of multicore scaling. In *Proceedings of the International Symposium on Computer Architecture* (pp. 365–376).
9. González, J., & González, A. (1997). Speculative execution via address prediction and data prefetching. In *Proceedings of the International Conference on Supercomputing* (pp. 196–203).
10. Hartstein, A., & Puzak, T. R. (2003). Optimum power/performance pipeline depth. In *Proceedings of the Annual IEEE/ACM International Symposium on Microarchitecture* (pp. 117–125).
11. Hennessy, J., Jouppi, N., Przybylski, S., Rowen, C., Gross, T., Baskett, F., et al. (1982). MIPS: A microprocessor architecture. In *Proceedings of the Annual Workshop on Microprogramming* (pp. 17–22).
12. Hennessy, J. L., & Patterson, D. A. (2012). *Computer architecture: A quantitative approach*, 5th ed. Cambridge: Morgan Kaufmann.
13. Ho, R., Mai, K. W., & Horowitz, M. A. (2001). The future of wires. *Proceedings of the IEEE, 89*(4), 490–504.
14. Hrishikesh, M. S., Burger, D., Jouppi, N. P., Keckler, S. W., Farkas, K. I., & Shivakumar, P. (2002). The optimal logic depth per pipeline stage is 6 to 8 fo4 inverter delays. In *Proceedings of the International Symposium on Computer Architecture* (pp. 14–24).
15. Hu, Z., Buyuktosunoglu, A., Srinivasan, V., Zyuban, V., Jacobson, H., & Bose, P. (2004). Microarchitectural techniques for power gating of execution units. In *Proceedings of the International Symposium on Low Power Electronics and Design* (pp. 32–37).
16. IBM. (1968). IBM System/360 Model 85 Functional Characteristics. Technical report, IBM Systems Reference Library. Second edition. Form A22-6916-1.
17. Koufaty, D., & Marr, D. T. (2003). Hyperthreading technology in the netburst microarchitecture. *IEEE Micro, 23*(2), 56–65.
18. Lamport, L. (1979). How to make a multiprocessor computer that correctly executes multiprocess programs. *IEEE Transactions on Computers, 28*(9), 690–691.
19. Marcuello, P., González, A., & Tubella, J. (1998). Speculative multithreaded processors. In *Proceedings of the International Conference on Supercomputing* (pp. 77–84).
20. McFarling, S. (1993). Combining branch predictors. Technical Report TN-36, Digital Western Research Laboratory.
21. Moore, G. (1965). Cramming more components onto integrated circuits. *Electronics, 38*(8), 114–117.
22. Moore, G. E. (1975). Progress in digital integrated electronics. *Technical Digest of the International Electron Devices Meeting, 13*, 11–12.
23. Patterson, D. A., & Sequin, C. H. (1981). RISC I: A reduced instruction set VLSI computer. In *Proceedings of the International Symposium on Computer Architecture* (pp. 443–457).
24. Pollack, F. Pollack's rule of thumb for microprocessor performance and area. http://en.wikipedia.org/wiki/Pollacks_Rule.
25. Radin, G. (1982). The 801 minicomputer. In *Proceedings of the International Symposium on Architectural Support for Programming Languages and Operating Systems* (pp. 39–47).
26. Smith, J. E. (1980). Branch predictor using random access memory. US Patent 4370711.
27. Sorin, D. J., Hill, M. D., & Wood, D. A. (2011). *A primer on memory consistency and cache coherence. Synthesis lectures on computer architecture*. San Rafael: Morgan & Claypool Publishers.
28. Sprangle, E., & Carmean, D. (2002). Increasing processor performance by implementing deeper pipelines. In *Proceedings of the International Symposium on Computer Architecture* (pp. 25–34).
29. Srinivasan, V., Brooks, D., Gschwind, M., Bose, P., Zyuban, V., Strenski, P. N., et al. (2002). Optimizing pipelines for power and performance. In *Proceedings of the Annual ACM/IEEE International Symposium on Microarchitecture* (pp. 333–344).
30. Thompto, B. (2016). Power9: Processor for the cognitive era. In *Hot Chips Symposium*.
31. Thornton, J. E. (1970). *Design of a computer: The Control Data 6600*. Glenview: Scott, Foresman and Company.

32. Tomasulo, R. M. (1967). An efficient algorithm for exploiting multiple arithmetic units. *IBM Journal of Research and Development, 11*(1), 25–33.
33. Tullsen, D. M., Eggers, S. J., & Levy, H. M. (1995). Simultaneous multithreading: Maximizing on-chip parallelism. In *Proceedings of the International Symposium on Computer Architecture* (pp. 392–403).
34. Wu, Q., Pedram, M., & Wu, X. (2000). Clock-gating and its application to low power design of sequential circuits. *IEEE Transactions on Circuits and Systems, 47*(3), 415–420.
35. Yeh, T.-Y., & Patt, Y. N. (1991). Two-level adaptive training branch prediction. In *Proceedings of the Annual International Symposium on Microarchitecture* (pp. 51–61).

Chapter 3
Aging Effects: From Physics to CAD

Hussam Amrouch, Heba Khdr, and Jörg Henkel

3.1 Introduction

Since the invention of integrated circuits, on-chip systems remarkably followed a continued trend of becoming denser every new generation towards gaining more advances in several aspects like performance, area, etc. However, this trend along with the relentless demand for reducing costs caused a noticeable decrease in the reliability due to various degradation effects. Even though the majority of degradation effects had been investigated for a long time, their impact had not reached a crucial point in which it would jeopardize the reliability of an entire on-chip system. In fact, an inflection point occurred when entering the nano-CMOS era. Technology roadmaps at that time provided several evidences that the aggressive scaling of technology nodes (i.e., ≤ 45 nm) would considerably increase the susceptibility of chips to varied kinds of failures [29]. This is because a transistor in such deep scaling reaches its limits and therefore the available design margins become insufficient to tolerate most of degradation effects. This, in particular, what makes current and upcoming on-chip systems constrained by reliability as maintaining the correct functionality of circuits during the entire projected lifetime cannot be easily compromised.

Download Software: This work is publicly available at [46] http://ces.itec.kit.edu/dependable-hardware.php.

H. Amrouch · H. Khdr · J. Henkel (✉)
Chair for Embedded Systems (CES), Karlsruhe Institute of Technology (KIT), Karlsruhe, Germany
e-mail: amrouch@kit.edu; heba.khdr@kit.edu; henkel@kit.edu

Degradation effects can be caused due to varied reasons either from the beginning during the design time or later during the operation time of circuits. To summarize, degradation effects can be divided into two main categories as follows [11]:

- *Time-Dependent Degradations:* These degradations are dependent on the chip's lifetime and they are mainly caused by aging phenomena. In practice, they slow down the switching speed of transistors, which makes circuits no longer reliably operate due to the increase in the likelihood of errors due to timing violations. Aging phenomena such as bias temperature instability (BTI) [25, 42, 55] and hot carrier-induced degradation (HCID) [41] are able to alter the key electrical characteristics of nMOS and pMOS transistors during runtime. The most observed ones are threshold voltage (V_{th}) and carrier mobility (μ) shifts.
- *Time-Independent Degradations:* They are independent of the chip's lifetime and mainly because of manufacturing variability at the design time as well as noise at the runtime. Manufacturing variability makes transistors with identical specifications have different electrical characteristics. Hence, they may unreliably operate at runtime. In addition, intrinsic noise such as random telegraph noise (RTN) and extrinsic noise such as energetic particles can randomly cause unpredictable fluctuations in the electrical characteristics of transistors. Based on the time and location (i.e., when and where such sudden changes occur), the reliability of circuit can be temporarily or permanently degraded.

Contrary to how they would appear at the first glance, the aforementioned degradation effects are actually interrelated with each other due to the existing *interdependencies* between them [4, 10, 11, 27]. As an example of such interdependencies is the relation between aging and the susceptibility to extrinsic noise (i.e., soft errors). While on the one hand aging degrades the electrical characteristics of transistors increasing the susceptibility of them to extrinsic noise [10], manufacturing variability, on the other hand, randomly influences the transistors and thus it makes them nonuniformly age over time.

In addition, BTI and HCID have strong interdependencies between each other because they share the same physical origin, which is interface traps. Similarly, BTI and RTN also have strong interdependencies between each other because they share the same physical origin, which is oxide traps. In Sect. 3.3, the interdependencies between degradation effects will be explained in detail.

3.1.1   Impact of Technology Scaling on Degradation Effects

Before reaching the 90-nm technology node, the same scaling factor for both supply voltage (S_V) and transistor length (S_L) was mainly used. This ensured that the electric fields remained almost constant [17] after scaling from a technology node to another, as shown in Eq. (3.1). This is widely known as Dennard's scaling [22]. However, reducing the geometry of transistors to integrate more and more transistors within a chip run counter to the need of maintaining a similar scaling factor. This

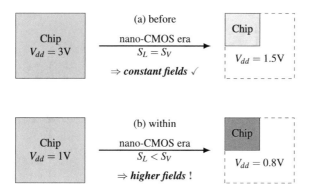

Fig. 3.1 Technology scaling results in constant electric fields when both area and voltage scale with the same factor. However, scaling within the nano-CMOS era results in higher fields as area and voltage are not scaled anymore with the same factor. Hence, more induced defects by aging and higher degradations will be caused due to the higher electric fields

is because V_{dd} scaling has almost reached its limit and since around a decade V_{dd} remained the same or has been reduced slightly with every new generation. In practice, this has changed the rule of scaling from ($S_L = S_V$) to ($S_L < S_V$), which made the electric fields across transistors become much stronger/higher (see Eq. (3.1) and Fig. 3.1 [2]). Similar to the electric field across the channel, the electric field over the gate has been also increased in smaller technology nodes. In fact, higher electric fields accelerate aging phenomena leading to higher degradations.

$$E_{new}(\text{Channel}) = \frac{V_{dd} \cdot S_V}{L \cdot S_L} \tag{3.1}$$

$$S_L = S_V \Rightarrow E_{new}(\text{Channel}) = E_{old}(\text{Channel})$$

$$S_L < S_V \Rightarrow E_{new}(\text{Channel}) > E_{old}(\text{Channel})$$

In addition, as technology nodes were being reduced to below 45 nm, the insulating quality of the gate dielectric became severely bad and thus leakage currents due to tunneling significantly increased. To cope this problem, manufactures replaced silicon dioxide, that has been for long time employed to form the gate dielectric, with other materials that have a higher dielectric constant such as hafnium-based dielectric layer from Intel [32], as shown in Fig. 3.2b. While the employment of such new materials was inevitable to enable the scaling of transistors, the susceptibility of transistors to aging effects became higher [18, 30] (e.g., HfO_2 exhibits less resiliency to aging effects than SiO_2). This is mainly because the new materials came with additional preexisting defects, which can be later activated during the operation of transistor. In short, scaling along with the high-K material has made both BTI and HCID become noticeable in both nMOS and pMOS transistors. This is against the traditional assumption of considering BTI in nMOS and HICD in pMOS to be negligible.

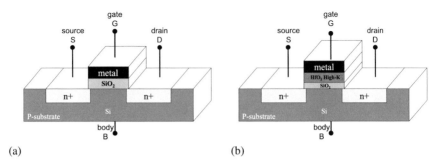

Fig. 3.2 Impact of scaling on a transistor. (**a**) An NMOS MOSFET without a high-K material. (**b**) The introduction of a high-K material

Finally, the discontinuation of Dennard's scaling has, additionally, led to increasing on-chip power densities and thus temperatures. Importantly, most of the underlying mechanisms of reliability degradations like BTI and HCID phenomena are stimulated by the operating temperature. This is because higher temperatures provide sufficient activation energy for defect generation at the physical level. Therefore, thermal management techniques at the system level (e.g., [23, 24]) are indispensable when it comes to reliability optimization. Such techniques are, in fact, needed to ensure that the thermal constraints are always fulfilled during runtime and thus reliability will be sustained during the entire projected lifetime. In general, in order to understand reliability we need to start the investigation from the physical level (where degradation effects originate) all the way up to the system level, where degradation effects finally take place (see Fig. 3.3).

> The continuation of Moore's law along with the discontinuation of Dennard's scaling has steadily increased the electric fields within transistors, which, in turn, has led to higher degradations.

3.2 Aging Effects at the Physical Level

Even though understanding the physical processes behind degradation effects is not entirely required at the circuit and system levels, there is a substantial need to analyze how they ultimately degrade the reliability [3, 5, 8]—*this holds even more when multiple degradation effects may interact with each other during the operation time of transistors.* As a matter of fact, degradation effects *jointly* occur within the gate dielectric of a transistor. Therefore, estimating correctly the overall impact of degradation effects on the electrical characteristics of transistors (e.g., V_{th}, μ, etc.) necessitates investigating them *jointly and separately.*

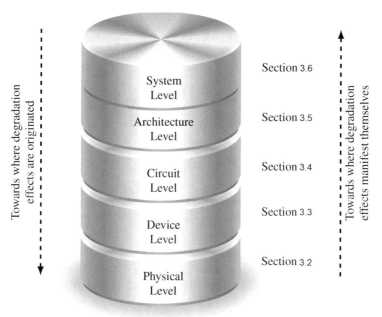

(The plotting code of this figure is obtained from [35])

Fig. 3.3 The abstraction levels, which are necessary to be involved to estimate the effects of aging

In the following, we will briefly explain the physical origins of the key degradation effects and then demonstrate the existing interdependencies between them.

(1) Bias Temperature Instability (BTI): It is the defect generation due to the vertical electric field when it is applied over the gate dielectric of a pMOS/nMOS transistor to form the conducting channel, as Fig. 3.4a [2] shows. The resulting defects are, according to the reaction-diffusion theory, mainly *interface traps* and *oxide traps*. In practice, the interface traps are formed when Si−H bonds break at the Si−SiO_2 interface and oxide traps are formed when charges are captured in the oxide vacancies within the dielectric [25]. During the operation of MOSFET and over time, these generated defects (i.e., interface/oxide traps) accumulate inside the transistor, manifesting themselves as degradations in its electrical characteristics (V_{th}, μ, etc.). Hence, aged transistors switch slower, which increases the likelihood of timing violations in circuits due to the operation at unsustainable clock frequencies. Further details are presented in Sect. 3.3.

In the following, we briefly summarize the process of interface trap generation due to BTI on a pMOS transistor that has H_2 as a diffusion medium [25].

- Due to the vertical electric field, a channel forms directly beneath the gate dielectric. In MOSFETs, the channel consists of minority charges, i.e., in a pMOS transistor, holes will be gathered near the gate dielectric.

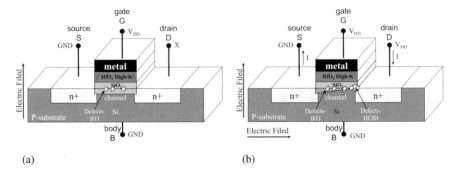

Fig. 3.4 Impact of scaling on a transistor. (**a**) Vertical electric field over the gate stimulates the BTI phenomenon leading to *evenly distributed* defects across the Si−SiO₂ interface. (**b**) Lateral electric field across the channel stimulates the HCID phenomenon leading to concentrated defects near to the drain

- Some of the carriers from the channel may recombine with electrons in the Si−H bond. In such a case, the Si−H bond, which had originally two electrons, has now only one electron left. This, in turn, weakens the bond and reduces the activation energy of both bonding partners.
- Through thermal activation, the H atom obtains sufficient kinetic energy to reach its activation energy. It bonds the remaining valence electron to its nucleus.
- The Si−H bond is broken and the remaining Si becomes positively charged, as its valence electron was taken by the H atom.
- The neutral H atom diffuses slowly through the SiO_2 grid within the gate dielectric through random hopping.
- If two neutral diffusing H atoms meet together, they bond into a molecular H_2, which is less reactive and thus it relatively diffuses faster through random hopping.
- The H_2 concentration increases, due to limited H_2 diffusion and therefore the $H + H \rightleftharpoons H_2$ reaction reaches an equilibrium state.
- Since the H concentration increases as the $Si−H + h^+ \rightleftharpoons Si \cdots H$ reaction saturates, the H_2 diffusion front moves through the gate dielectric. As soon as the H_2 diffusion front reaches the poly-crystalline Si, H_2 diffuses away deep into the system (i.e., almost lost forever).
- H_2 concentration reaches an equilibrium as diffusion out of the system and H_2 generation finds a steady state.

Interface traps:

$$Si−H + h^+ \rightleftharpoons Si^+ + H$$

Molecular hydrogen:

$$H + H \rightleftharpoons H_2$$

Interface trap generation:

$$Si-H \xrightarrow{h+} Si\cdots H$$

$$Si\cdots H \longrightarrow Si^+ + H$$

Interface trap recovery:

$$Si^+ + H \longrightarrow Si\cdots H$$

$$Si\cdots H \longrightarrow Si-H + h^+$$

It is noteworthy that recent studies [54, 56] demonstrated that the BTI-induced degradation (ΔV_{th}) can be observed within a short time domain (e.g., 1 µs) in the advanced technology nodes. The reason of why this has not been observed before is the lack of advanced equipments to conduct ultra-fast BTI measurements [56]. This, in fact, is a paradigm shift in aging from a sole long-term reliability degradation, as in the traditional view, to short reliability degradation.

(2) Hot Carrier-Induced Degradation (HCID): It is the defect generation near the drain terminal of transistor due to *hot carriers* whose high kinetic energies are sufficiently significant to form electron–hole pairs through ionization. Unlike BTI, which is caused by the vertical electric field over the gate, HCID is caused by the lateral electric field across the channel of a transistor [33]. It is noteworthy that "*hot*" does not refer here to the temperature of transistor but, instead, to the ability of carriers to tunnel out of the semiconductor material and then get injected in undesirable areas of the transistor. Similar to BTI, HCID, over time, also causes charge buildup near the drain of a transistor (see Fig. 3.4b). Thus, it results in degrading the electrical characteristics of transistors (similar to BTI) [40].

In the following, we briefly summarize the process of interface trap generation due to BTI on a pMOS transistor that has H_2 as a diffusion medium.

- HCID occurs when both vertical and later electric fields are applied. The first one is in charge of forming the transistor's channel and the latter is in charge of moving the carriers within the channel from the source to the drain.
- During the movement of carriers within the channel (i.e., holes in the case of pMOS), some of them may gain sufficient kinetic energy through the acceleration in which the activation energy of the H atom in the Si−H bond is reached.
- Due to the vertical electric field and Rutherford scattering, the holes are hitting the interface between the channel and the gate dielectric. In practice, if a "hot" hole, i.e., a hole with a high kinetic energy, hits an H atom, it is activated due to the deposition of the kinetic energy upon the H atom.
- The Si−H bond is broken due to the kinetic activation and the remaining Si becomes positively charged as its electron recombines with the remaining hole. The neutral H atom diffuses into the gate dielectric. Then, similar to the BTI mechanism, the H may diffuse away from the system.

Aging Recovery Mechanism: Healing the broken Si−H bonds is the reverse process to the breaking of the Si−H bonds. Breaking and healing are exact opposites and the damage due to breaking of the bonds can be almost entirely recovered by the healing process, if sufficient relaxation time is given. A full recovery is impossible due to the irreversible nature of the removal of H_2 from the system (according to the reaction–diffusion theory of aging). Once H_2 reaches the poly-crystalline gate, it is forever lost and the corresponding interface traps cannot be healed, due to the lack of a bonding partner.

Healing follows this reaction equation:

$$2\,Si^+ + H_2 \longrightarrow 2\,(Si{-}H) + 2\,h^+$$

Recovery of Si−H bonds is independent of how the bonds were broken and thus exactly the same for BTI and HCID. In the following, we give a brief explanation on how the recovery can happen in a pMOS transistor.

1. Reaction of $H_2 \longrightarrow H + H$.
2. Diffusion of atomic H through random hopping.
3. An atomic H atom finds a reactive Si with an unsatisfied bond, i.e., a missing electron.
4. Both partners form a bond with the reaction: $Si^+ + H \longrightarrow Si{-}H + h^+$.
5. Holes tunnel through gate dielectric either towards the channel or more likely (due to the vertical electric field) through the gate dielectric towards the poly-crystalline gate electrode.

Most broken bonds might heal almost immediately, as the atomic H is still near the unsatisfied Si^+. It is also worthy to note that the healing process reduces the H-concentration, which will accelerate the degradation at other places due to the $Si^+ + H \rightleftharpoons Si{-}H + h^+$ equilibrium.

(3) Random Telegraph Noise (RTN) It is probabilistic capturing/emitting of carriers in oxide traps [53], which is partially due to quantum effects that cause carriers to tunnel from the channel deep into the gate dielectric charging an oxide trap. Such captured carriers within the dielectric lead to undesirable charges which, in practice, interact with the vertical electric field. Therefore, they manifest themselves as an increase in V_{th} [50].

(4) Random Dopant Fluctuation (RDF) It is because of randomness in the dopant concentration during the fabrication process of MOSFET. RDF directly leads to fluctuations in the channel's formation of MOSFETs during the operation. Hence, RDF influences the V_{th} of MOSFETs [14] and makes it follow a statistical distribution. In the deep nanoscale (e.g., 22 nm), RDF effects become more severe as the dopant concentrations increase, narrowing the tolerance for the doping process or increasing the variation for a fixed tolerance. At the same time, smaller geometries reduce the number of dopants despite higher concentrations resulting in a significant impact on the transistor V_{th} even for a slight change in the absolute number of dopants [14].

(5) Process Variation (PV) It is due to the fluctuations in the geometry of transistor such as in its length (L) and width (W) during the fabrication process. Such geometric variance results in fluctuations in MOSFET's electrical characteristics during its operation because the drain current of MOSFET, which mainly determines its switching speed, depends on both W and L. To demonstrate that Eq. (3.2) shows the first order of approximation of the I_D equation in MOSFET transistors. Note that the following equation represents an abstracted and simplified expression for the drain current in transistor and in current technology nodes (where short channel effects are important) the drain current equation is significantly more complex.

$$\text{Delay} = \frac{1}{I_D} \; ; \; I_D \approx \frac{\mu}{2} \cdot C_{\text{ox}} \cdot \frac{W}{L} \cdot (V_{\text{dd}} - V_{\text{th}})^2 \tag{3.2}$$

3.2.1   Interdependencies of Degradation Effects

On the one hand, when operating a MOSFET transistor both of vertical and lateral electric fields may be generated (see Fig. 3.4b). Thus, both BTI and HCID phenomena will be *jointly* occurring. As explained earlier, both BTI and HCID share the same kind of defects which is *interface traps*. Hence, it cannot be excluded that they may amplify or cancel each other during operation time. Hence, modeling BTI and HCID needs to be done jointly and not separately. Otherwise, the overall impact is either over- or underestimated [10, 11].

On the other hand, manufacturing variability through PV and RDF results in random variations of the electrical properties of transistors. While PV varies the length (L) and width (W) of transistors, RDF varies the dopant concentration in the substrate near the channel of MOSFETs. Due to the inherent interdependencies between the key parameters of MOSFET, the aforementioned fluctuations in V_{th}, W, and L lead to other fluctuations in the, for example, I_D, transconductance (g_m), etc. [11].

In [10], we studied the interdependencies of BTI and HCID and showed how the overall degradation of V_{th} and μ is a function of the total number of defects present as follows:

$$\Delta V_{\text{th}}(t+1) = \frac{q}{C_{\text{ox}}} \cdot (\Delta N_{\text{IT}}(t) + \Delta N_{\text{OT}}(t))$$

$$\Delta \mu(t+1) = \frac{\mu(t_0)}{1 + \alpha \cdot \Delta N_{\text{IT}}(t)}$$

$$\Delta N_{\text{IT}}(t+1) = \Delta N_{\text{IT.BTI}}(t) + \Delta N_{\text{IT.HCID}}(t)$$

Here, $\Delta N_{\text{IT.BTI}}(t)$ and $\Delta N_{\text{IT.HCID}}(t)$ are the number of induced *interface traps* at time t due to BTI and HICD, respectively. $\Delta N_{\text{OT}}(t)$ is the number of *oxide traps* due to BTI. Over time, this leads to an interdependency between BTI and HCID since the amount of induced defects by each is recursively dependent on the current V_{th}, as

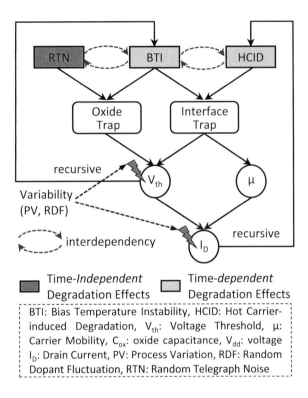

Fig. 3.5 The shared interface traps between both BTI and HCID creates an interdependency between both phenomena over time [11]. Similarly, the shared oxide traps create also an interdependency between both RTN and BTI over time. In addition, the effects of process variation cannot be excluded from the existing interdependencies

it can be observed in Fig. 3.5. For a full description of the defect modeling of N_{IT}, N_{HT}, and N_{OT}, please refer to the original work [10]. In this work, we employed physics-based aging models based on [1, 31].

Importantly, the initial values of parameters at the beginning (i.e., $V_{\mathrm{th}}(t_0)$ and $\mu(t_0)$) vary from a transistor to another due to manufacturing variability (RDF and PV). In other words, aging-induced degradations depend on the manufacturing variability as the latter randomly alters the initial state of the electrical properties of transistors.

To model the impact of PV effects, a normal distribution can be typically employed:

$$f(x, \mu, \sigma) = \frac{1}{\sigma\sqrt{2\pi}} \cdot e^{-\frac{(x-\mu)^2}{2\sigma^2}} \tag{3.3}$$

with μ as the mean and σ as the standard deviation for the transistor's geometries (i.e., L and W). Regarding RDF, we model the V_{th} distribution according to [14] as follows:

$$\sigma_{V_{\mathrm{th}}} = \frac{q \cdot t_{\mathrm{ox}}}{\epsilon_{\mathrm{ox}}} \cdot \sqrt{\frac{N_A \cdot W_d}{4 \cdot L_{\mathrm{eff}} \cdot W_{\mathrm{eff}}}} \tag{3.4}$$

Then, the properties of each transistor in the circuit exhibit probabilistic distributions. The changes in V_{th} and μ influence other parameters due to the interdependencies between them. The latter is intrinsically inside the MOSFET compact modeling [49]. In our work, we employ Berkeley short-channel IGFET model (BSIM), which is the industrial compact model of MOSFET [49].

Last but not least, the other kind of defects caused by BTI, which is oxide traps, is shared between RTN. This also creates interdependencies between the BTI and RTN phenomena, similar to the interdependencies between the BTI and HCID phenomena, as it is demonstrated in Fig. 3.5. However, it is noteworthy that RTN exhibits a random behavior due to the inherent nature of carrier tunneling, while BTI deterministically behaves as it is driven by the applied electric field over the gate dielectric. In general, the overall degradation induced by RTN is, typically, defined by the total number of currently activated traps and the sum of induced ΔV_{th} by each of these traps. This necessitates employing two distributions in order to model the total impact of RTN on the ΔV_{th}. The first distribution, obtained from [45], describes the distribution of traps due to RTN in current technologies. Whereas, the second distribution is based on the analysis in [50] that describes the induced ΔV_{th} per trap. As a result, based on these two employed distributions, we can model the random impact of RTN on V_{th} as follows:

$$\Delta V_{th}(RTN) = X \cdot Y \quad : \quad \text{where}$$

$$X \in f(X, \mu_{traps}, \sigma_{traps}) = \frac{1}{\sigma_{traps}\sqrt{2\pi}} \cdot e^{-\frac{(X-\mu_{traps})^2}{2 \cdot \sigma_{traps}^2}}$$

$$Y \in f(Y, \mu_{V_{th}}, \sigma_{V_{th}}) = \frac{1}{\sigma_{V_{th}}\sqrt{2\pi}} \cdot e^{-\frac{(Y-\mu_{V_{th}})^2}{2 \cdot \sigma_{V_{th}}^2}}$$

Because of the shown analysis of RTN [45], which we rely on in our modeling, has been performed for transistors of $L_{ref} = 25$ nm and $W_{ref} = 45$ nm, we augment the RTN modeling in Eq. (3.5) with scaling factors to allow proper modeling in which other transistors' sizes are taken into consideration. To achieve that, we include in Eq. (3.5) the factor $S = \frac{L_{ref} \cdot W_{ref}}{L_{eff} \cdot W_{eff}}$, because RTN depends on $\frac{1}{L_{eff} \cdot W_{eff}}$ [51]. Thus, RTN can be modeled as:

$$\Delta V_{th}(RTN) = X \cdot Y \cdot \frac{L_{ref} \cdot W_{ref}}{L_{eff} \cdot W_{eff}} \tag{3.5}$$

The Induced Interdependency between BTI and RTN As BTI generates more *oxide traps* over time, the distribution of X gets shifted towards higher numbers and therefore RTN gains more strength. [20] demonstrated that the ΔV_{th} due to RTN doubles after BTI stress, so over a typical lifetime of, e.g., 10 years, the μ_{traps} can be shifted to twice its value following the typical power law of $\frac{1}{6}$ of BTI [31].

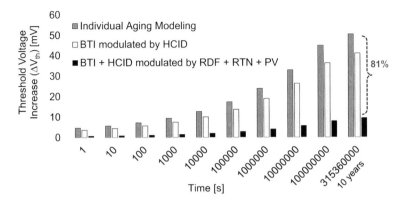

Fig. 3.6 Impact of considering the existing interdependencies between different degradation effects on estimating the overall impact quantified by the threshold voltage increase (ΔV_{th})

In Fig. 3.6, we show a comparison with respect to the estimated ΔV_{th} when degradation effects are *separated* and *jointly* considered. As shown, neglecting the existing interdependencies leads to higher estimations. As a result, a more conservative (i.e., *inefficient*) guardband, i.e., employing larger guardbands than what is actually needed, needs to be later employed to protect against aging effects. As shown in Fig. 3.6, the consideration of HCID jointly with BTI leads to a smaller guardband (8%) compared to the consideration of BTI alone. In addition, considering other degradation effects leads to further reduction to 81%. As explained, this is because that degradation effects can influence each other. This shows why it is indispensable to consider degradation effects *jointly* and not separately in order to estimate the overall impact of aging correctly.

3.3 Aging Effects at the Device Level

Generated interface traps due to BTI and HCID as well as generated oxide traps due to BTI results in different shifts in the key parameters of MOSFETs. The most observable one is an increase in the threshold voltage of transistor (V_{th}). However, aging-induced defects also degrade other electrical characteristics of MOSFETs like carrier mobility (μ), subthreshold slope (SS), and gate–drain capacitance (C_{gd}) [9].

(1) **Threshold Voltage Increase** (V_{th}): As charge builds up at the $SiO_2 - Si$ interface due to interface traps, the electric field becomes weaker and weaker due to the interactions with the accumulated positively charged Si^+ atoms. The weaker the electric field, the less minority carriers will be attacked when forming the transistor's channel. This manifests itself as an increase in the threshold voltage (V_{th}) of transistor and, in fact, a higher gate voltage (V_G) is

necessary to obtain the same electric field and maintain the original (i.e., before aging) threshold voltage (V_{th_0}).

(2) Carrier Mobility Decrease (μ): Due to the positive charge buildup at the SiO_2-Si interface, the moving carriers within the transistor's channel encounter more resiliency when flowing from the source to the drain of transistor. This is mainly because these carriers and the generated interface traps, which are positive charges, will be repelling each other due to Coulomb scattering. In other words, the positively charged Si^+-ions act as an obstacle and carrier mobility sinks due to the higher resistance against movement.

It is noteworthy that the activated oxide traps also build up positive charges within the gate dielectric. However, in contrast to the previously discussed defects (i.e., interface traps), these charges are not near the SiO_2-Si interface but are located deep within the gate dielectric. Therefore, they are not able to hinder the mobility of carriers in the channel, but only affect the electric field over the gate dielectric. Similar to the interface traps, oxide traps also weaken the formation of the transistor's channel and thus they also manifest themselves as an increase in the threshold voltage (V_{th}) of transistor.

(3) Subthreshold slope Degradation (SS): Analogous to μ, only interface traps (N_{IT}) contributes to SS degradation. This is because N_{IT} are donor type traps at the Si/IL interface which are positively charged when empty of electrons (or filled with holes) and neutral when filled with electrons [9]. This nature of the traps being filled or emptied in response to the gate voltage gives rise to an interface trap capacitance (C_{IT}) which comes in parallel with the depletion capacitance and hence it contributes to SS degradation. Note that oxide traps also have a negligible contribution to SS degradations because such traps are not close to the interface and thus they cannot interact with the carriers inside the transistor's channel.

(4) Gate–Drain Capacitance Increase (C_{gd}): As discussed, generated interface trap will perform an intrinsic capacitance which will result in a distortion in the capacitance voltage (CV) characteristics of the transistor [37]. Among all the capacitances of MOSFETs, gate–drain capacitance (C_{gd}) change plays the most important role because the resulting impact is magnified due to Miller effect at the output drain node [9]. Hence, both the delay and the dynamic power are directly influenced as the transistor will face a larger capacitance to charge.

3.4 Aging Effects at the Circuit Level

In our work [5], we studied the impact that ΔV_{th} has on the delay of logic gates. We demonstrated that the same amount of degradation *disproportionally* alters the delay of gates within a standard cell library. In addition, we also showed that aging-induced ΔV_{th} may even improve (instead of degrade) the delay of some gates under specific operating conditions $OPCs$ (i.e., input signal slew and output load capacitance). Based on the aforementioned observations, we proposed

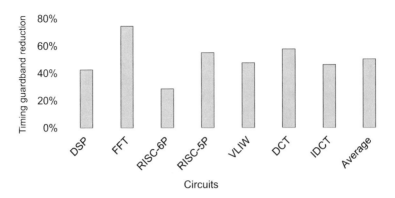

Fig. 3.7 Impact of employing our degradation-aware cell libraries during circuit synthesis on *containing* timing guardbands

to provide the synthesis tools with our so-called degradation-aware cell library. Such a library contains the detailed delay information of every gate but under the effects that worst-case aging degradations (e.g., the maximum ΔV_{th}) have. When employing such a degradation-aware cell library during circuit synthesis, the mature algorithms in synthesis tool flows—evolved over decades in commercial EDA software like Synopsys and Cadence—will automatically select gates which exhibit less degradation under aging (i.e., gates with smaller delay). Hence, aging can be proactively at the design time considered and circuits with a better performance can be obtained. To illustrate the impact of aging-aware synthesis, we show in Fig. 3.7 the saving in timing guardband for different complex circuits. In this analysis, we used the Synopsys Processor Designer Tool to generate the RTL for different kinds of processor designs such as VLIW and RISC processors. As it can be observed, our aging-aware logic synthesis leads to 51% smaller guardbands on average and up to $\sim$75%. Analogously, we have also employed the idea of performing logic synthesis in the scope of temperature, instead of aging effects. Our analysis showed an 18%, on average, and up to 24% saving in temperature guardbands [7] for different processors. This demonstrates the efficiency of bringing awareness of degradation effects to the existing logic synthesis tool flows to employ their mature optimization algorithms towards containing timing guardbands.

Our degradation-aware cell libraries are publicly available at [46] allowing other researchers to employ them. They are ready to use with existing tool flows (e.g., Synopsys) without requiring any modifications.

We also applied the concept of degradation-aware cell libraries in order to evaluate the final impact of considering the existing interdependencies between degradation effects. As demonstrated earlier in the previous section (see Fig. 3.6), neglecting the interdependencies leads to overestimating the overall degradation w.r.t ΔV_{th}. To evaluate further and to translate the results from ΔV_{th} into the resulting delay increase of circuits, we created the corresponding cell libraries under the different estimated ΔV_{th} cases. We summarize the results of the impact

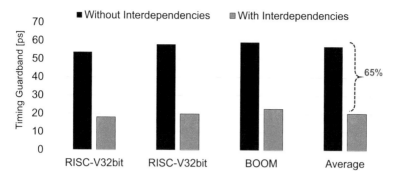

Fig. 3.8 Comparison between considering interdependencies (i.e., BTI, HCID, RTN, RDF and, PV *jointly*) and neglecting interdependencies (i.e., BTI alone) on the resulting timing guardbands of different processors. Considering interdependencies allows *reducing the guardbands* by 65% on average

of considering the interdependencies on guardbands in Fig. 3.8. In this evaluation, we studied two different kinds of in-order processors, namely: 32-bit and 64-bit. Both of them are based on the open-source instruction set architecture RISC-V [13]. In addition, we also studied the Berkeley Out-of-Order Machine (BOOM) which is state-of-the-art industry-competitive RISC-V processor [19]. The RTL descriptions of the three processors were generated using the so-called Rocket Chip Generator [12]. As it can be seen in Fig. 3.8, considering the existing interdependencies between degradation effects leads to 64% smaller guardbands on average.

3.5 Aging-Induced Approximation

As discussed earlier, aging effects increase the delay of gates within circuits and hence the delay of critical paths prolong during the lifetime. This, in turn, will lead to timing violations if no sufficient timing guardbands is included. However, timing guardbands mean, in practice, operating the circuits at lower frequency than its potential and therefore a considerable efficiency loss will be incurred. In this section, we demonstrate first whether we really need to include timing guardbands even for circuits that can inherently tolerate errors like image processing. This is necessary to explore how much aging effects can degrade the quality of service in such fault-tolerant circuits.

To achieve that, we study the discrete cosine transform (DCT) and inverse discrete cosine transform (IDCT) circuits as they are typically employed in multimedia designs to encode and decode images, respectively. We used our degradation-aware cell libraries (detailed in Sect. 3.4) in order to extract the precise delay information of every gate within the netlists of DCT-IDCT circuits. Then, we perform gate-

Original Image [20]   **Year 1** (Balance-case)   **Year 10** (Balance-case)

Fig. 3.9 Analyzing the impact of aging-induced timing errors on the quality of image. Aging effects, when no timing guardbands in included, lead to a quality loss even after merely 1 year [5]

level simulations using Mentor ModelSim in which the DCT-IDCT chain encodes and then decodes an input image. Then, by computing the peak signal-to-noise ratio(PSNR) of the output image, we quantify how the aging degrades the quality of image. Note that in the gate-level simulations, the baseline clock of the DCT-IDCT chain, which is the clock of the circuit in the absence of aging, is used. Hence, errors will occur due to timing violations as no timing guardband is included. This enables us to evaluate whether we really need to include a timing guardband in circuits that can tolerate errors like image processing circuits. Note that the PSNR is a typical metric used to represent image quality and for human the value 30 is considered the minimum accepted quality [52].

In Fig. 3.9 [5], we show the original image before aging and the resulting image after 1 year of aging stress as well as after 10 years lifetime. As can be noticed, even in image processing circuits, which can tolerate errors, aging effects can lead to a severe quality loss which is not accepted even after merely 1 year of operation. This clearly demonstrates the necessity of including timing guardbands.

In order to completely remove timing guardbands while still maintaining quality, we proposed in [6], for the first time, the concept of aging-induced approximation. Because aging effects lead to timing violations, which are stochastic and uncontrolled, we explored how approximate computing principles can be employed in order to translate aging effects into deterministic and controlled approximations. In Fig. 3.10, we demonstrate an example of our characterization of a 32×32 bit adder. In this analysis, we consider both worst- and actual-case aging. For the latter, we use stimuli from a normal distribution as well as inputs extracted from an IDCT decoding a sample input image. As shown, reducing the precision of the adder enables us to compensate the aging-induced delay increase. For instance, reducing the precision by merely 2 bits allows us to *narrow* the required guardband by 31%. Importantly, reducing the precision further down to 24 bits will be sufficient to *completely remove* the guardband until 1 year of worst-case aging. Precision needs to be further reduced down to 22 bits for a reliable operation (i.e., without any aging-induced timing errors) over 10 years without adding any guardband. However, considering actual-case (instead of worst-case) aging provides us with a

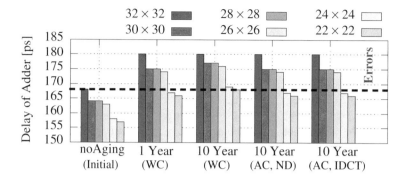

Fig. 3.10 Characterization of a 32 × 32 bit adder to translate aging-induced timing errors into an equivalent precision reduction. Here, WC: *worst-case* aging. (AC, ND): *actual-case* aging under inputs from a *normal distribution*. (AC, IDCT): *actual-case* aging under IDCT inputs of an input sample image [6]

Fig. 3.11 Examples of images when applying our aging-induced approximations. The lifetime of 10 years under the worst-case aging has been considered in this analysis [6]

less conservative precision reduction of only 8 bits. Note, however, that considering actual case of aging will not provide a guarantee that timing errors will never occur during the projected lifetime.

To examine how translating stochastic timing errors caused by aging into deterministic and controlled approximations can impact the image quality, we show in Fig. 3.11 some examples of the output images obtained from the IDCT circuit. The examined images for this analysis were extracted from the "video trace library" [57], which is a standard benchmark for multimedia analysis. A lifetime of 10 years under worst-case aging was used to determine approximations and hence presents the lower boundary of PSNR. As can be noticed, PSNR on average merely drops by 8db and it is above 30db in all images except one ("mobile"), where it is slightly below. Examples of obtained images are presented in Fig. 3.11. As shown, even for the image in the case of "mobile" with 28db PSNR, the quality is still very good and noise is hardly observable.

3.6   Aging at the System Level

As discussed in the previous sections, aging effects have become a major reliability concern. Although they originate at physical level, aging effects have interdependency with system-level parameters [34–36].

Figure 3.12 illustrates this interdependency between physical- and system-level parameters. In particular, system level means like increasing number of parallel threads of the applications and upscaling the voltage and frequency level of the cores lead to increase the number of active cores and the application performance, and that, in turn, elevates the temperature on the chip. Increasing the temperature and the supply voltage (V_{dd}) induce and accelerate aging mechanisms, which manifest themselves as increase in the key transistor's parameters like threshold voltage (ΔV_{th}), as discussed earlier. Such degradations in the electrical characteristics of transistors prolong the critical path delay of the processor, and eventually timing errors might occur, which of course degrade the system reliability.

This interdependency between system-level parameters and physical-level parameters motivates us to investigate the possibility of tackling aging concern at the system level. Since the ultimate goal of system-level resource management is performance optimization, we highlight in the following subsections two system-level potentials to optimize the performance while considering aging.

3.6.1   Aging-Aware Performance Optimization

As illustrated in Fig. 3.12, the shift in the threshold voltage, i.e., ΔV_{th}, is increased by increasing temperature and V_{dd} [25]. Figure 3.13 illustrates the resulting increase in V_{th} along with V_{dd} and temperature. These values are obtained using a physics-based aging model. As mentioned above, these two factors are resulting from the decisions taken at the system level.

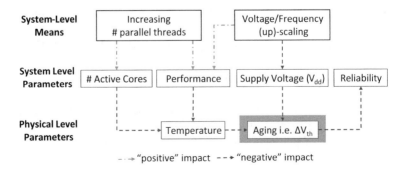

Fig. 3.12 The interdependency between system-level parameters and physical-level parameters

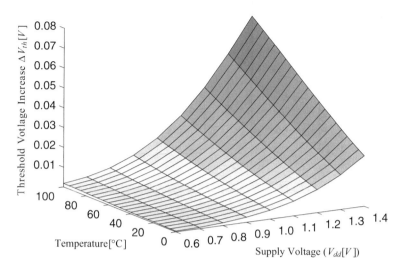

Fig. 3.13 The resulting increase in the threshold voltage along with temperature and voltage [34]

To sustain reliability regardless of system-level decisions, chip designers add to the clock delay a timing guardband that compensates the worst-case increase in the threshold voltage ΔV_{th}^{worst}, as seen in Fig. 3.14. ΔV_{th}^{worst} is determined through measuring the processor delay increase after exposing a prototyping chip to accelerated aging conditions [32]. This is, however, pessimistic and hinders the system from using the existing capabilities of the hardware. Instead of that, employing a guardband based on more realistic aging constraint (e.g., ΔV_{th}^{max} in Fig. 3.14) would be more efficient. However, in this case, the aging constraint ΔV_{th}^{max} must be enforced at the system level. In practice, ΔV_{th}^{max} indicates the maximum expected degradation in the threshold voltage for the entire targeted lifetime in the processor. Thus, resource management must make its decisions for performance optimization under aging constraint ΔV_{th}^{max} [34].

The majority of the state-of-the-art works that consider aging at the system level do not aim at maximizing the performance under an aging constraint, but instead they aim at minimizing or balancing aging to increase the lifetime (e.g., [21, 44, 47, 48]). Few recent works have proposed resource management techniques that aim at maximizing the performance under aging constraint. In [26], a resource management technique is proposed to enforce the aging constraint at runtime. Namely, when the aging constraint is violated, the voltage and frequency (v/f) levels of the cores are downscaled in order to decrease temperatures and thus aging. The drawback of this technique is that it does not consider the positive impact of reducing the V_{dd} on aging [25], i.e., the lower the V_{dd}, the less aging is induced. That means, unnecessary v/f level downscaling is applied leading to unnecessary performance losses. Contrarily, in [43] the impact of both, the temperature and V_{dd}, on aging is considered. To meet the aging constraint, the proposed resource

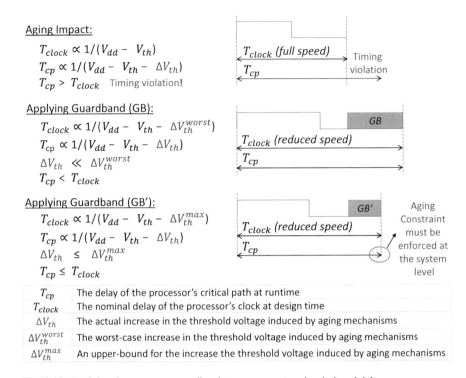

Aging Impact:

$$T_{clock} \propto 1/(V_{dd} - V_{th})$$
$$T_{cp} \propto 1/(V_{dd} - V_{th} - \Delta V_{th})$$
$$T_{cp} > T_{clock} \quad \text{Timing violation!}$$

Applying Guardband (GB):

$$T_{clock} \propto 1/(V_{dd} - V_{th} - \Delta V_{th}^{worst})$$
$$T_{cp} \propto 1/(V_{dd} - V_{th} - \Delta V_{th})$$
$$\Delta V_{th} \ll \Delta V_{th}^{worst}$$
$$T_{cp} < T_{clock}$$

Applying Guardband (GB'):

$$T_{clock} \propto 1/(V_{dd} - V_{th} - \Delta V_{th}^{max})$$
$$T_{cp} \propto 1/(V_{dd} - V_{th} - \Delta V_{th})$$
$$\Delta V_{th} \leq \Delta V_{th}^{max}$$
$$T_{cp} \leq T_{clock}$$

T_{cp}	The delay of the processor's critical path at runtime
T_{clock}	The nominal delay of the processor's clock at design time
ΔV_{th}	The actual increase in the threshold voltage induced by aging mechanisms
ΔV_{th}^{worst}	The worst-case increase in the threshold voltage induced by aging mechanisms
ΔV_{th}^{max}	An upper-bound for the increase the threshold voltage induced by aging mechanisms

Fig. 3.14 Applying the necessary guardbands to compensate aging-induced delay

management in [43] employs conservative voltage and the temperature bounds within its decision-making process. However, this approach is conservative, because it does not exploit the full available design space for performance optimization, as will be discussed later on.

To achieve the goal of maximizing performance under aging constraint, there is a need to explore the design space that satisfies the predetermined aging constraint for the targeted lifetime. For example, Fig. 3.15 quantifies ΔV_{th} after 1 year. The three cutoff lines represent specific aging constraints for the targeted lifetime. The design space of temperature and V_{dd} below each of these cutoff lines represents feasible options. Figure 3.16 shows the feasible design space considering one of these constraints, i.e., $\Delta V_{th} = 0.03V$. The first observation is that this (red) cutoff line can be translated into a *set of temperature constraints*, each corresponds to a specific V_{dd} such that the aging constraint is satisfied. For instance, if V_{dd} is set to 1.25 V, the temperature constraint will be equal to 52 °C. However, if a lower V_{dd} is selected, e.g., 1.15 V, the temperature can reach 86 °C without violating the aging constraint. As mentioned above, the state-of-the-art technique [43] enforces one pair of temperature and V_{dd} bounds. For instance, if the temperature constraint is 60 °C, the maximum allowed V_{dd} is 1.225 V. The two dashed lines in Fig. 3.16 represent an exemplary pair of temperature and V_{dd} bounds employed by the state

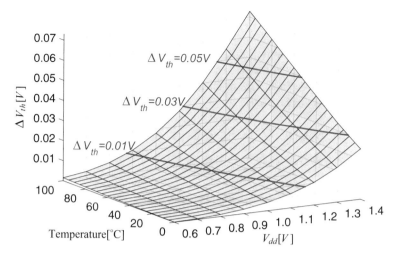

Fig. 3.15 Aging-aware design space: accurate (based on physical models) inter-dependencies between *amount of aging* (i.e., ΔV_{th}), temperature, and voltage V_{dd} [34]

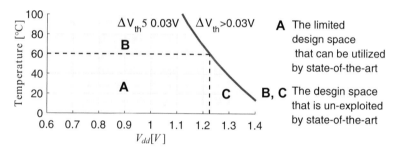

Fig. 3.16 The *entire* design space that satisfies the aging constraint and the *limited* design space resulting from applying conservative temperature and V_{dd} bounds (dashed area) by state-of-the-art [34]

of the art. Therefore, the resource management of the state of the art cannot, by principle, select temperature and V_{dd} outside this limited space, while considering the entire aging-aware design space enables new optimization potential.

> It is indispensable to fully exploit the entire aging-aware design space by resource management techniques in order to optimize for performance under aging constraint.

3.6.2 Guardband-Aware Performance Optimization

As aforementioned, sustaining reliability necessitates employing a guardband to compensate the slowdown that arises as transistors age. The reason of this slowdown is the reduction in the drain current (I_d) as it is seen from the first-order approximation of I_d in Eq. (3.6).

$$\text{After aging: } I_d \approx (V_{dd} - V_{th} - \Delta V_{th}) \tag{3.6}$$

$$\text{Before aging: } I_d \approx (V_{dd} - V_{th})$$

$$\text{Transistor (delay)} \propto \frac{1}{I_d} \Rightarrow \text{ aged transistors are slower}$$

There are two main *guardband types* that can be used to compensate this aging-induced delay, they are frequency guardband and voltage guardband [36]:

(1) **Frequency Guardband (F_GB):** To prevent timing violations, the frequency needs to be reduced by a certain amount such that the clock delay becomes always longer than the critical path delay. In practice, an additional time slack ($t_{\text{guardband}}$), corresponding to the maximum aging-induced delay for the targeted lifetime must be added to the nominal critical path delay (t_{nom}), as Eq. (3.7) shows. Hence, the resulting frequency $f_{\text{guardband}}$ will be lower than the nominal frequency f_{nom}.

$$f_{\text{guardband}} = \frac{1}{t_{\text{nom}} + t_{\text{guardband}}} < f_{\text{nom}} = \frac{1}{t_{\text{nom}}} \tag{3.7}$$

(2) **Voltage Guardband (V_GB):** As shown in Eq. (3.6), the drain current of the transistor I_d will be reduced due to aging. To compensate this reduction in the I_d, a guardband can be added to the nominal voltage (V_{nom}), as Eq. (3.8) shows. Hence, the processor can still be clocked with the nominal frequency without exhibiting timing errors even though aging effects may occur during its lifetime. However, the new voltage $V_{\text{guardband}}$ will be higher than the nominal voltage V_{nom}.

$$V_{\text{guardband}} = V_{\text{nom}} + \Delta V_{th} > V_{\text{nom}} \tag{3.8}$$

Typically, either frequency guardband (F_GB) or voltage guardband (V_GB) is selected at design time of the processor (circuit level) and adopted throughout its lifetime. The guardband type is selected based on system requirements. In particular, V_GB enables using the nominal frequency without any reduction. Therefore, it is selected when the performance of the targeted system cannot be scarified. V_GB, however, increases the power consumption. Therefore, F_GB is selected for low-power systems, because it does not increase the power. However, F_GB decreases the performance.

Our investigation in the following case study shows that *the ultimate impact of a guardband type cannot be accurately estimated at the circuit level irrespective from the workload at the system level.*

Motivational Case Study This case study shows the impact of the guardband type on the power consumption and the performance for two different applications, i.e., "canneal" and "x264," from the PARSEC benchmark suite [15], running on 64-core chip simulated by Gem5 [16] and McPAT [39]. We run two experiments for each application to test V_GB and F_GB. In each experiment, eight instances of the application are executed on the cores, with 8 parallel threads for each. The nominal voltage V_{nom} and the nominal frequency f_{nom} are 1.4 V and 4 GHz, respectively. The estimated $V_{guardband}$ and $f_{guardband}$ according to the employed physics-based aging model are 1.491 V and 3.42 GHz, respectively. We assume that the targeted multicore system employs a thermal management unit (TMU), that is responsible to avoid thermal violations by power-gating the cores that exceed the predefined critical temperature ($T_{crit} = 100\,°C$) of the chip.

The power consumption will increase when applying V_GB, since it leads to increasing the supply voltage as seen in Eq. (3.8), and thereby increasing the power consumption of the core (see Eq. (3.9)).

$$P = \alpha \cdot C_{eff} \cdot V_{dd}^2 \cdot f + P_{leak}\,(V_{dd}, T) \tag{3.9}$$

where α indicates the activity factor of the core and C_{eff} indicates the effective switching capacitance of the core. P_{leak} is the leakage power, which depends on the supply voltage and the core's temperature. According to our simulation using McPAT and considering the worst-case temperature on the chip (T_{crit}), P_{leak} is equal to 3.5 W and 4.2 W for V_{nom} and $V_{guardband}$, respectively. As an example, in the case of "x264," α and C_{eff} are equal to 0.3 and 2.2, respectively. We use Eq. (3.9) for (V_{nom}, $f_{guardband}$) and for ($V_{guardband}$, f_{nom}) to calculate the resulting power consumptions of applying F_GB and V_GB, respectively. That results in total power consumptions of 8 W and 10 W, as shown in Fig. 3.17. As expected, V_GB results in higher power consumption compared to F_GB.

The resulting performance when applying V_GB is increased compared to F_GB in the case of "canneal," which is expected because F_GB leads to lower frequency (see Eq. (3.7)) and thereby lower processor speed. However, in the case of "x264," applying V_GB reduces the performance with 24% compared to F_GB. The reason is that the higher increase in the power consumption in the case of "x264," when V_GB is employed, triggers the TMU which power-gates additional 14 cores whose estimated temperatures exceed T_{crit}. Consequently, the system performance is degraded in this case. This is unlike the case of "canneal" in which the increase in the power consumption due to employing V_GB does not lead to thermal violations. As a result, V_GB is not always able to provide better performance than F_GB despite the higher operating frequency [36].

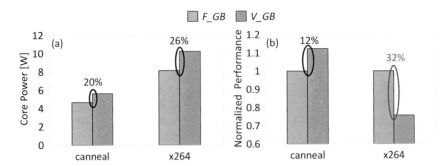

Fig. 3.17 Even though V_GB allows operating at a higher frequency compared to F_GB, it does not always provide a better performance in a thermally constrained system. For x264, employing V_GB leads to less performance, due to the incurred thermal violations [36]. (**a**) shows the total consumed power. (**b**) shows the normalized performance when applying frequency and voltage guardbands

Therefore, traversing from the circuit level to the system level is necessary to be capable of considering the workload-induced temperatures and thereby accurately estimating the impact of the guardband type on performance. That enables selecting the best fitting guardband type w.r.t. the workload, with the goal to maximize the performance while sustaining reliability.

Acknowledgements This work is supported in part by the German Research Foundation (DFG) as part of the priority program "Dependable Embedded Systems" [28] (SPP 1500—spp1500.itec.kit.edu). Authors would like to thank Victor van Santen from CES, KIT for his very valuable work and contribution as well as our research partners; Souvik Mahapatra, Subrat Mishra from IIT Bombay, Andreas Gerstlauer from University of Texas at Austin, and Montserrat Nafria and Javier Martin-Martinez from Universitat Autonoma de Barcelona.

References

1. Alam, M., Roy, K., & Augustine, C. (2011). Reliability- and process-variation aware design of integrated circuits. *IEEE International Reliability Physics Symposium (IRPS)* (pp. 4A.1.1–4A.1.11).
2. Amrouch, H. (2015). *Techniques for aging, soft errors and temperature to increase the reliability of embedded on-chip systems.* PhD thesis, Dissertation, Karlsruhe Institute of Technology (KIT).
3. Amrouch, H., Ebi, T., & Henkel, J. (2013). Stress balancing to mitigate NBTI effects in register files. In *2013 43rd Annual IEEE/IFIP International Conference on Dependable Systems and Networks (DSN)* (pp. 1–10).
4. Amrouch, H., & Henke, J. (2017). Containing guardbands. In *2017 22nd Asia and South Pacific Design Automation Conference (ASP-DAC)* (pp. 537–542).

5. Amrouch, H., Khaleghi, B., Gerstlauerz, A., & Henkel, J. (2016). Reliability-aware design to suppress aging. In *Proceedings of the 53rd Annual Design Automation Conference* (p. 12).
6. Amrouch, H., Khaleghi, B., Gerstlauer, A., & Henkel, J. (2017, June). Towards aging-induced approximations. In *2017 54th ACM/EDAC/IEEE Design Automation Conference (DAC)* (pp. 1–6).
7. Amrouch, H., Khaleghi, B., & Henkel, J. (2017). Optimizing temperature guardbands. In *Design, Automation & Test in Europe Conference & Exhibition (DATE)*.
8. Amrouch, H., Martin-Martinez, J., van Santen, V., Moras, M., Nafria, M., Rodriguez, R., et al. (April 2015). Connecting the physical and application level towards grasping aging effects. In *2015 IEEE International Conference on Reliability Physics Symposium (IRPS)* (pp. 3D. 1.1–3D. 1.8).
9. Amrouch, H., Mishra, S., van Santen, V., Mahapatra, S., & Henkel, J. (2017). Impact of BTI on dynamic and static power: From the physical to circuit level. In *2017 IEEE International Reliability Physics Symposium (IRPS)* (pp. CR–3). New York: IEEE.
10. Amrouch, H., van Santen, V., Ebi, T., Wenzel, V., & Henkel, J. (2014, November). Towards interdependencies of aging mechanisms. In *2014 IEEE/ACM International Conference on Computer-Aided Design (ICCAD)* (pp. 478–485).
11. Amrouch, H., van Santen, V. M., & Henkel, J. (2016). Interdependencies of degradation effects and their impact on computing. *IEEE Design & Test*.
12. Asanovi, K., Avizienis, R., Bachrach, J., Beamer, S., Biancolin, D., Celio, C., et al. (2016). The rocket chip generator. Technical Report No. UCB/EECS-2016-17.
13. Asanovi, K., & Patterson, D. A. (2014, August). Instruction sets should be free: The case for RISC-v. Technical Report UCB/EECS-2014-146, EECS Department, University of California, Berkeley.
14. Asenov, A. (1998). Random dopant induced threshold voltage lowering and fluctuations in sub-0.1 μm MOSFET's: A 3-D ldquo;atomistic rdquo; simulation study. *IEEE Transactions on Electron Devices, 45*, 2505–2513.
15. Bienia, C., Kumar, S., Singh, J. P., & Li, K. (2008). The PARSEC benchmark suite: Characterization and architectural implications. In *Proceedings of the 17th International Conference on Parallel Architectures and Compilation Techniques (PACT)* (pp. 72–81).
16. Binkert, N., Beckmann, B., Black, G., Reinhardt, S. K., Saidi, A., Basu, A., et al. (2011). The gem5 simulator. *ACM SIGARCH Computer Architecture News, 39*(2), 1–7.
17. Borkar, S. (1999). Design challenges of technology scaling. *IEEE Micro, 19*(4), 23–29.
18. Cartier, E., Kerber, A., Ando, T., Frank, M., Choi, K., Krishnan, S., et al. (2011, December). Fundamental aspects of HfO$_2$-based high-k metal gate stack reliability and implications on t$_{inv}$-scaling. In *2011 IEEE International Electron Devices Meeting (IEDM)* (pp. 18.4.1–18.4.4).
19. Celio, C., Patterson, D. A., & Asanovic, K. (2015). The Berkeley out-of-order machine (boom): An industry-competitive, synthesizable, parameterized risc-v processor. Technical Report UCB/EECS-2015-167, EECS Department, University of California, Berkeley.
20. Chiu, J., Chung, Y., Wang, T., Chen, M.-C., Lu, C., & Yu, K. (2012). A comparative study of NBTI and RTN amplitude distributions in high gate dielectric pMOSFETs. *IEEE Electron Device Letters, 33*(2), 176–178.
21. Das, A., Shafik, R. A., Merrett, G. V., Al-Hashimi, B. M., Kumar, A., & Veeravalli, B. (2014). Reinforcement learning-based inter- and intra-application thermal optimization for lifetime improvement of multicore systems. In *Proceedings of the 51st Annual Design Automation Conference*, DAC '14 (pp. 170:1–170:6). New York: ACM.
22. Dennard, R., Rideout, V., Bassous, E., & LeBlanc, A. (1974). Design of ion-implanted MOSFET's with very small physical dimensions. *IEEE Journal of Solid-State Circuits, 9*(5), 256–268.
23. Ebi, T., Al Faruque, M. A., & Henkel, J. (2009). Tape: Thermal-aware agent-based power economy for multi/many-core architectures. In *Proceedings of the 2009 International Conference on Computer-Aided Design*, ICCAD '09 (pp. 302–309).

24. Ebi, T., Kramer, D., Karl, W., & Henkel, J. (2011, October). Economic learning for thermal-aware power budgeting in many-core architectures. In *2011 Proceedings of the Ninth IEEE/ACM/IFIP International Conference on Hardware/Software Codesign and System Synthesis (CODES+ISSS)* (pp. 189–196).
25. Goel, N., Naphade, T., & Mahapatra, S. (2015). Combined trap generation and transient trap occupancy model for time evolution of NBTI during DC multi-cycle and AC stress. In *2015 IEEE International Reliability Physics Symposium (IRPS)* (pp. 4A.3.1–4A.3.7).
26. Haghbayan, M. H., Miele, A., Rahmani, A. M., Liljeberg, P., & Tenhunen, H. (2017). Performance/reliability-aware resource management for many-cores in dark silicon era. *IEEE Transactions on Computers, 66*(9), 1599–1612.
27. Henkel, J., & Amrouch, H. (2016). Designing reliable, yet energy-efficient guardbands. In *2016 IEEE International Conference on Electronics, Circuits and Systems (ICECS)* (pp. 540–543). New York: IEEE.
28. Henkel, J., Bauer, L., Becker, J., Bringmann, O., Brinkschulte, U., Chakraborty, S., et al. (2011). Design and architectures for dependable embedded systems. In *CODES+ISSS* (pp. 69–78).
29. International technology roadmap for semiconductors (ITRS). http://www.itrs.net.
30. Joshi, K., Hung, S., Mukhopadhyay, S., Chaudhary, V., Nanaware, N., Rajamohnan, B., et al., (2013, April). HKMG process impact on N, P BTI: Role of thermal IL scaling, IL/HK integration and post HK nitridation. In *2013 IEEE International Reliability Physics Symposium (IRPS)* (pp. 4C.2.1–4C.2.10).
31. Joshi, K., Mukhopadhyay, S., Goel, N., & Mahapatra, S. (2012). A consistent physical framework for N and P BTI in HKMG MOSFETs. In *2012 IEEE International Reliability Physics Symposium (IRPS)* (pp. 5A.3.1–5A.3.10).
32. Keane, J., & Kim, C. H. (2011). Transistor aging. *IEEE Spectrum, 48*(5), 28–33.
33. Khan, S., & Hamdioui, S. (2010). Trends and challenges of SRAM reliability in the nano-scale era. In *5th International Conference on Design and Technology of Integrated Systems in Nanoscale Era* (pp. 1–6).
34. Khdr, H., Amrouch, H., & Henkel, J. (2018). Aging-constrained performance optimization for multi cores. In *Proceedings of the 55th Annual Design Automation Conference* (p. 63). New York: ACM.
35. Khdr, H., Amrouch, H., & Henkel, J. (2018). Aging-aware boosting. *IEEE Transactions on Computers.*
36. Khdr, H., Amrouch, H., & Henkel, J. (2018). Dynamic guardband selection: Thermal-aware optimization for unreliable multi-core systems. *IEEE Transactions on Computers.*
37. Krishnan, A. T., Reddy, V., Chakravarthi, S., Rodriguez, J., John, S., & Krishnan, S. (2003). NBTI impact on transistor and circuit: models, mechanisms and scaling effects [MOSFETs]. In *IEEE International Electron Devices Meeting, 2003. IEDM'03 Technical Digest* (pp. 14–15).
38. Latex plotting code. https://github.com/MartinThoma/LaTeX-examples/
39. Li, S., Ahn, J.-H., Strong, R., Brockman, J., Tullsen, D., & Jouppi, N. (2009). McPAT: An integrated power, area, and timing modeling framework for multicore and manycore architectures. In *42nd IEEE/ACM International Symposium on Microarchitecture (MICRO)* (pp. 469–480).
40. Mahapatra, S., Saha, D., Varghese, D., & Kumar, P. B. (2006). On the generation and recovery of interface traps in MOSFETs subjected to NBTI, FN, and HCI stress. *IEEE Transactions on Electron Devices, 53*(7), 1583–1592.
41. Maricau, E., & Gielen, G. (2011). Computer-aided analog circuit design for reliability in nanometer CMOS. *IEEE Journal on Emerging and Selected Topics in Circuits and Systems, 1*(1), 50–58.
42. Martin-Martinez, J., Kaczer, B., Toledano-Luque, M., Rodriguez, R., Nafria, M., Aymerich, X., et al. (2011, April). Probabilistic defect occupancy model for NBTI. In *2011 IEEE International Reliability Physics Symposium (IRPS)* (pp. XT.4.1–XT.4.6).

43. Mercati, P., Paterna, F., Bartolini, A., Benini, L., & Rosing, T. (2017, September). Warm: Workload-aware reliability management in Linux/Android. *IEEE Transactions on Computer-Aided Design of Integrated Circuits and Systems, 36*(9), 1557–1570.
44. Mück, T. R., Ghaderi, Z., Dutt, N. D., & Bozorgzadeh, E. (2017). Exploiting heterogeneity for aging-aware load balancing in mobile platforms. *IEEE Transactions on Multi-Scale Computing Systems, 3*(1), 25–35.
45. Nagumo, T., Takeuchi, K., Yokogawa, S., Imai, K., & Hayashi, Y. (2009). New analysis methods for comprehensive understanding of random telegraph noise. In *2009 IEEE International Electron Devices Meeting (IEDM)* (pp. 1–4). New York: IEEE.
46. Our released models, tools and degradation-aware cell libraries. http://ces.itec.kit.edu/dependable-hardware.php
47. Raparti, V. Y., Kapadia, N., & Pasricha, S. (2017, April). ARTEMIS: An aging-aware runtime application mapping framework for 3D NoC-based chip multiprocessors. *IEEE Transactions on Multi-Scale Computing Systems, 3*(2), 72–85.
48. Rathore, V., Chaturvedi, V., & Srikanthan, T. (2016). Performance constraint-aware task mapping to optimize lifetime reliability of manycore systems. In *Proceedings of the 26th Edition on Great Lakes Symposium on VLSI*, GLSVLSI '16 (pp. 377–380). New York: ACM.
49. Singh Chauhan, Y., Venugopalan, S., Paydavosi, N., Kushwaha, P., Jandhyala, S. Duarte, J. P., et al. (2013). BSIM Compact MOSFET Models for SPICE Simulation. In *2013 Proceedings of the 20th International Conference Mixed Design of Integrated Circuits and Systems (MIXDES)* (pp. 23–28).
50. Tega, N., Miki, H., & Pagette, F., et. al. (2009). Increasing threshold voltage variation due to random telegraph noise in FETs as gate lengths scale to 20 nm. In *VLSIT*.
51. Tega, N., Miki, H., Yamaoka, M., Kume, H., Mine, T., Ishida, T., et al. (2008). Impact of threshold voltage fluctuation due to random telegraph noise on scaled-down SRAM. In *IEEE International Reliability Physics Symposium, 2008. IRPS 2008* (pp. 541–546).
52. Thomos, N., Boulgouris, N., & Strintzis, M. (2006). Optimized transmission of JPEG2000 streams over wireless channels. *IEEE Transactions on Image Processing, 15*(1), 54–67.
53. Tsukamoto, Y., Toh, S. O., Shin, C., Mairena, A., Liu, T.-J. K., & Nikolic, B. (2010). Analysis of the relationship between random telegraph signal and negative bias temperature instability. In *IRPS* (pp. 1117–1121).
54. van Santen, V. M., Amrouch, H., Martin-Martinez, J., Nafria, M., & Henkel, J. (2016). Designing guardbands for instantaneous aging effects. In *Proceedings of the 53rd Annual Design Automation Conference* (p. 69).
55. van Santen, V. M., Amrouch, H., Parihar, N., Mahapatra, S., et al. (2016). Aging-aware voltage scaling. In *Design, Automation & Test in Europe Conference & Exhibition (DATE)* (pp. 576–581).
56. van Santen, V. M., Martin-Martinez, J., Amrouch, H., Nafria, M. M., & Henkel, J. (2018). Reliability in super-and near-threshold computing: A unified model of RTN, BTI, and PV. *IEEE Transactions on Circuits and Systems I: Regular Papers, 65*(1), 293–306.
57. Video trace library. http://trace.eas.asu.edu/yuv/index.html

Part II
OS Layer

Chapter 4
The HARPA-OS

Simone Libutti, Giuseppe Massari, and William Fornaciari

4.1 Reliability-Aware Run-Time Resource Management

The HARPA-OS acts at software level. In fact, it employs task scheduling—hence the mapping of computing resources—as a software knob to tackle faults, aging, and performance variability.

Indeed, this kind of approach has already been proved to be effective. Huang et al. [7], for instance, propose a task scheduling policy that is based on a reliability model, and whose objective is to maximize the system lifetime. The policy requires a power model of the computing resources; however, such model is rarely available. Most importantly, the policy does not take into account the entity of the degradation that is currently experienced by the hardware.

The approach presented by Chou and Marculescu [4] addresses the latter shortcoming. In fact, they take into account permanent, transient, and intermittent faults of the computing cores. Their resource management strategy specifically targets NoC-based platforms, and the approach is based on the presence of spare cores that can be employed when those that are currently being used experience faults.

Sun et al. [11] propose an approach that also targets NoC-based platforms. It is based on a negative-bias temperature instability (NBTI) [10] model. In this case, the resource mapping policy is based on linear programming, which is known to be poorly performing at runtime. Similar considerations can be done for the work presented in [6].

Haghbayan et al. [5] present a more structured approach where a reliability management unit provides a prediction about the current aging status of the

S. Libutti (✉) · G. Massari · W. Fornaciari
Politecnico di Milano, Milano, Italy
e-mail: simone.libutti@polimi.it; giuseppe.massari@polimi.it; william.fornaciari@polimi.it

© Springer International Publishing AG, part of Springer Nature 2019   73
W. Fornaciari, D. Soudris (eds.), *Harnessing Performance Variability
in Embedded and High-performance Many/Multi-core Platforms*,
https://doi.org/10.1007/978-3-319-91962-1_4

hardware resources. A mapping unit is in charge of allocating tasks considering the boundaries defined by the reliability management unit and some geometrical constraints.

The most complete solution is probably the WARM framework presented by Marcati et al. [9]. The framework relies on a reliability model [2] and on multiple controllers that act at different time granularity to set the reference temperature, perform task allocation, and set voltage/frequency points for each core. The framework has been implemented on an Android-based mobile platform. Although we consider WARM an interesting solution, our approach is based on the idea of optimizing resource usage by monitoring the runtime behavior of applications. Moreover, our approach targets a very wide range of architectures from high-tier embedded to HPC; therefore, our needs are best addressed by a more scalable approach.

4.2   Managing Resources from User-Space

A resource manager is a software layer that orchestrates configuration and allocation of computational resources while taking into account system-wide and user-specific goals. Relying on a resource manager instead of distributing the management logic throughout the entire software stack is a very convenient approach for both operating systems and applications developers, since it allows them to move most of the management complexity in a black box that is easily portable and maintainable. That is why, during the last years, there was a strong push towards migrating the resource management logic into user-space processes [13].

Figure 4.1 shows a typical example of user-space-managed environment. Unmanaged applications run, as usual, on top of the operating system, and they are allowed to use a predefined set of general purpose resources. Managed applications run instead on top of the resource manager, which, according to some optimization policy, computes which is the best set of resources that must be allocated to each application. In order to monitor the system and to enforce the allocation choices, the resource manager relies, respectively, on the monitors and the knobs that are exposed by the operating system.

It is worth noticing that some resource managers move the enforcement complexity to the applications side. That is, the resource manager monitors the resources and computes allocations, while applications are in charge of enforcing the allocation, e.g., by using thread affinity to pin their threads on the allocated cores.

The HARPA-OS is a user-space daemon that dynamically allocates computing resources to managed applications (from now on, only "applications"). With "dynamic allocation," we refer to the ability of changing the resource allocation during runtime in order to address changes in system status, resource availability, or applications' Quality of Service (QoS) requirements. One of the most interesting features of the HARPA-OS is that, conversely to the typical job schedulers, it provides applications with an explicitly resource-aware and quality-aware execution

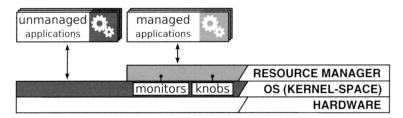

Fig. 4.1 A user-space-managed environment. Unmanaged applications run on top of the operating system, while managed applications run on top of the resource manager, which exploits the monitors and knobs exposed by the operating system for management purposes

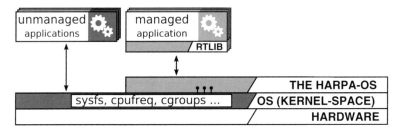

Fig. 4.2 A HARPA-OS-managed environment. Unmanaged applications run on top of the operating system, while managed applications rely on the HARPA-OS Runtime Library, which transparently negotiates resource allocation with the HARPA-OS

flow: the processing of applications is in fact divided into bursts whose quality and resource demand are individually monitored. The allocation may change only between processing bursts; in that case, applications are notified about the new allocation so that they are able to reconfigure their own software parameters (e.g., number of threads) accordingly before the upcoming burst.

Figure 4.2 shows a typical example of HARPA-OS-managed environment. Applications do not directly communicate with the resource manager: they instead rely on the HARPA-OS Runtime Library (*RTLib*), which transparently negotiates resource allocation with the HARPA-OS. In order to monitor the system and to enforce resource allocation, the HARPA-OS in turn relies on the available monitors and knobs exposed by the operating system, e.g., sysfs, cpufreq, and the Linux Control Groups.

4.3   The Abstract Execution Model

To synchronize their processing with the dynamical resource allocation, the HARPA-OS requires applications to relinquish the control of their execution flow to the *RTLib*. That is, the *RTLib* drives the execution of running applications by choosing when to execute processing bursts and reconfiguration routines. To benefit from such support, the applications' code must be moved into a C++ class that exposes the following methods:

setup   Setting up the processing, e.g., spawning threads and performing mallocs;
configure   (re) Configuring the application software parameters according to the current allocation;
run   Process the next chunk of data;
monitor   Monitor the current QoS. If needed, ask for a higher or lower one;
release   Terminate the application, e.g., join threads and free mallocs.

During runtime, the aforementioned methods will be invoked by the *RTLib* according to the HARPA-OS resource and quality-aware execution flow shown in Fig. 4.3. When the application has started, the *RTLib* asks for resources to the HARPA-OS (yellow arrow) and, while waiting for the allocation to be computed,

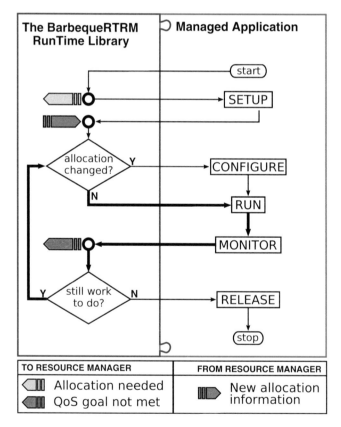

Fig. 4.3 The HARPA-OS: execution flow of a managed application. The runtime library drives the execution and transparently synchronizes it with the runtime-variable resource allocation. Bold arrows indicate the most common execution flow, i.e., a continuous burst of run and monitor invocations under a constant resource allocation

it invokes the application *setup* method, which will take place in a predefined set of generic (shared) resources. While the application performs the setup, the HARPA-OS allocates it some resources and notifies the decision to the *RTLib* (green arrow). After the *setup*, the *RTLib* can therefore invoke the *configure* method, where the application, being aware of the current allocation, is able to configure itself accordingly. After that, the *RTLib* lets the application fall into the common execution flow (bold pattern), which consists bursts of *run* and *monitor* methods that are executed under a constant resource allocation. An allocation change can be triggered either by external events (e.g., some resources become unavailable) or in case of an unsatisfactory QoS. In the latter case, the *RTLib* notifies the problem to the *HARPA-OS* (orange arrow), which will correct the allocation as soon as possible while letting the application continue its execution bursts.

4.4 Integrating Applications

In this subsection, we provide an example of application integration. Listing 4.1 shows the simplified code of a multimedia unmanaged application that processes a stream of video frames. Once started, the application performs an initialization and selects a parallelism level equal to the number of online cores (lines 3–7). Then, it processes the frames in bursts of one frame per active thread (lines 12–21). Finally, it checks the result and terminates (lines 23–24).

Listing 4.2 shows the main function of the corresponding integrated application. The original code is moved to a class, i.e., `ManagedTask`, that links with the RTLib. As a consequence, the main function is simplified: it just initializes the RTLib and the `ManagedTask` class (lines 4–5), then it starts the processing and waits for it to terminate (lines 8–9).

Listing 4.3, instead, shows the implementation of the C++ class that encapsulates the application code by defining the Setup, Configure, Run, Monitor, and Release methods. Apart from the logic that was added to perform runtime reconfiguration and to check if the current QoS is satisfactory, the resulting code is the same to that of the original application.

4.5 Defining Resource Allocations

Given that the execution flow of managed applications is based on the concepts of QoS compliance and software parameters reconfiguration, applications are supposed to be profiled at design time in order to compute a set of static allocations, each of which leads to a known QoS level and needs a known set of software parameters to be appropriately exploited. Accordingly, the HARPA-OS scheduling policies base their allocation choices on a static set of applications' configurations, which are called application working modes (AWMs).

Listing 4.1 Simplified example of a frame processing application in its nonintegrated version. The application parses the arguments, initializes some data structure, and selects a parallelism level. Then, it processes the frames in "threads_number"-sized bursts. Finally, it checks the results, joins the threads, and terminates

```
1  void main(int argc, char *argv[])
2  {
3      ParseCommandLine(argc, argv);
4      // Load data, initialize structures, ...
5      Initialize();
6      // Using one thread per core (constant!)
7      SetThreadsNumber(sysconf(_SC_NPROCESSORS_ONLN));
8
9      int result;
10     int current_frame = 0;
11
12     while(HaveWorkToDo(current_frame)) {
13         // Process one frame per thread and update frame count
14         result = ProcessFrames(current_frame);
15
16         // Handle processing errors, if any
17         if (result) {
18             HandleError(result);
19             break;
20         }
21     }
22
23     CheckResult();
24     JoinThreads();
25 }
```

Listing 4.2 Main file of the integrated version of the application from Listing 4.1. The application initializes the RTLib and uses it to instantiate the class whose methods encapsulate the processing code. Then, it launches the processing and waits for it to terminate

```
1  void main(int argc, char *argv[])
2  {
3      // Initialize the RTLib
4      rtlib rtlib_services;
5      ManagedTask t(rtlib_services, argc, argv);
6
7      // Start the application
8      t->StartExecution();
9      t->WaitTermination();
10 }
```

Relying on a static set of configurations is a reasonable choice, since, in order to be managed by the HARPA-OS, applications must be nonetheless refactored. This, in turn, means that the typical managed workload is entirely composed of applications that are known and can therefore profiled at design time. However, relying on static configurations also leads to complex issues:

• Profiles that are obtained at design time may be inaccurate at runtime, e.g., due to different frequencies, temperature variation, aging-induced performance variability, data variability, or presence of concurrently running applications;

Listing 4.3 Simplified code of the integrated version of the application from Listing 4.1. Underlined functions are calls to the RTLib APIs

```
1  ////// DEFINITION OF THE MANAGEABLE CLASS //////
2  class ManagedTask: public ManagedApplication
3  {
4      public:
5      ManagedTask(rtlib rtlib_services, int argc, char *argv[]) {
6          ParseCommandLine(argc, argv);
7      };
8
9      int Setup() override;
10     int Configure() override;
11     int Run() override;
12     int Monitor() override;
13     int Release() override;
14
15     int result;
16     int current_frame = 0;
17  };
18
19  ////// IMPLEMENTATION  OF THE MANAGEABLE CLASS //////
20  int ManagedTask::Setup() {
21      Initialize();
22      return RTLIB_OK;
23  }
24
25  int ManagedTask::Configure() {
26      // Using one thread per allocated core (runtime-variable!)
27      SetThreadsNumber(get_alloc_cores());
28      return RTLIB_OK;
29  }
30
31  int ManagedTask::Run() {
32      if (! HaveWorkToDo(current_frame))
33          return RTLIB_WORKLOAD_NONE;
34
35      // Process one frame per thread and update frame count
36      result = ProcessFrames(current_frame);
37
38      if (result) {
39          HandleError(result);
40          return RTLIB_ERROR;
41      }
42      return RTLIB_OK;
43  }
44
45  int ManagedTask::Monitor() {
46      if (DontLike(get_throughput())) complain();
47      return RTLIB_OK;
48  }
49
50  int ManagedTask::Release() {
51      CheckResult();
52      JoinThreads();
53      return RTLIB_OK;
54  }
```

- An exhaustive application profiling (i.e., profiling the application for each software and parameter combination and for each possible resource allocation) is complex and time wasting. Moreover, it must be carried out once for each different architecture;

Listing 4.4 Example of application recipe. A recipe is an XML file that contains a static list of application working modes (AWMs). When computing the resource allocation for an application, the scheduling policy chooses an AWM between those that are listed in its recipe

```
1  <?xml version="1.0"?>
2  <HARPA-OS recipe_version="0.8">
3  <!-- Priority with respect to other managed applications.
4       The lower, the better (minimum 1). -->
5  <application priority="1">
6  <!-- Known allocations for any Linux/CGroups-based multi-core -->
7  <platform id="org.linux.cgroup">
8      <awms>
9          <!-- AWM 0: low quality (since it got the lowest value) -->
10         <awm id="0" name="only_one_cpu" value="10">
11             <resources>
12                 <!-- Allocation: a whole core -->
13                 <!-- Note: resource IDs do not matter, they are
14                      used internally for registration purposes -->
15                 <cpu id="0">
16                     <pe qty="100"/>
17                 </cpu>
18             </resources>
19         </awm>
20         <!-- AWM 1: high quality (x10 with respect to AWM 0) -->
21         <awm id="1" name="cpu_and_gpu" value="100">
22             <resources>
23                 <!-- Allocation: 20% of a core -->
24                 <cpu id="0">
25                     <pe qty="20"/>
26                 </cpu>
27                 <!-- Allocation: a GPU -->
28                 <gpu id="0">
29                     <pe qty="100"/>
30                 </gpu>
31             </resources>
32         </awm>
33     </awms>
34 </platform>
35 </application>
36 </HARPA-OS>
```

- Discrete allocations may lead to instability. For instance, if the current resource demand of an application is three cores but the profiled configurations feature two or four cores, the allocation will periodically switch between the two available allocations.

It is worth noticing that HARPA-OS also supports scheduling policies that compute allocations in the continuous space. That is, resource allocations are computed on-the-fly with disregard of the profiled configurations. We will deal with continuous allocation later in this chapter.

Listing 4.4 shows an example of application recipe, which is an XML file that contains the static list of AWMs. When computing how many and which resources will be allocated to an application, the scheduling policy chooses an AWM between those that are listed in the corresponding recipe and tries to wisely mapping it on the available resources. The recipe can contain multiple platform sections, each of whom lists the AWMs for a particular architecture. Each AWM is identified by an ID, a human-readable description, a Quality of Service level (expressed as an integer value), and the list of resources that must be allocated to the application in order to make it reach (ideally) that QoS level.

4.6 HARPA-OS Scheduling Policies

The HARPA-OS features a modular structure that allows it to be easily adapted to the user needs. Regarding the hardware, HARPA-OS runs on any Linux-based systems; moreover, it can be integrated with the runtime of hardware bare-metal accelerators. Regarding system-wide management, HARPA-OS bases resource allocation choices on scheduling policies, which are implemented as plug-ins and make use of the HARPA-OS internal APIs to assess the current system and applications' status and to actuate scheduling decisions.

The structure of HARPA-OS is shown in Fig. 4.4. Yellow boxes indicate components that are implemented as plug-ins and can therefore be easily swapped to adapt the HARPA-OS behavior to the user needs. The resource manager interacts with hardware and applications through an application and a platform proxy, respectively. Communication with applications is based on a communication channel, whose default implementation is based on remote procedure calls (RPC). The heart of the resource manager is the scheduling policy, which is also implemented as a plug-in.

In the context of the HARPA project, we implemented two HARPA-OS policies. The first one is *TEMPURA*, which targets the minimization of power consumption in embedded systems. *TEMPURA* addresses the problem of bounding the system power consumption according to a given budget, as required by the two embedded system (ES) HARPA use cases. For instance, in the high-end ES use case (Spectrum Sensing application on Freescale I.MX6Q platform by THALES) we need to set a power budget to enable the deployment of such a class of applications on mobile devices. In the low-end ES use case (Beesper system based on the ODROID-XU3 platform by HENESIS), instead, we must guarantee the system operation under conditions of energy budget variability, and we must prevent the overheating of the processing resources that would turn the system off. *TEMPURA* allows HARPA-OS to pursue a twofold objective: it strives in keeping the temperature of the processor below a critical threshold, while guaranteeing a target uptime given the current energy budget.

The second scheduling policy is *PerDeTemp*, which instead counteracts performance variability and minimizes temperature in HPC systems. *PerDeTemp* aims at consolidating the computing resources utilization, and it can be used in both embedded and HPC scenarios. It was developed as a more powerful version of YaMS, which is the default scheduling policy in HARPA-OS and is based on the already mentioned applications' recipes.

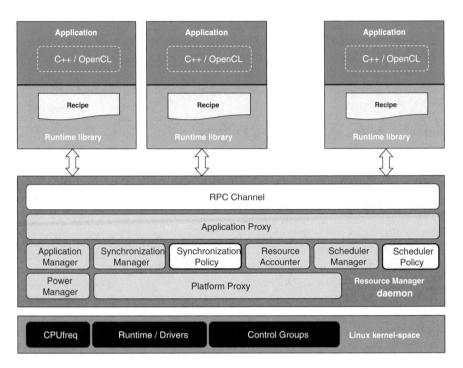

Fig. 4.4 Structure of the HARPA-OS. Yellow boxes indicate components that are implemented as plug-ins and can therefore be easily swapped to adapt the HARPA-OS behavior to the user needs. The resource manager (red layer) interacts with hardware and applications through an application and a platform proxy, respectively. The heart of the resource manager is the scheduling policy, which is also implemented as a plug-in

Conversely to YaMS, *PerDeTemp* is recipe-less, i.e., resource allocations are computed on-the-fly with disregard of the profiled configurations. Recipe-less scheduling enables two main benefits: a) the scheduling policy can handle also completely unknown applications and is robust to changes of datasets or architecture; *b)* allocations are not in the discrete space anymore, and this allows the scheduling policy to converge to optimal allocations instead of continuously switching between near-optimal ones. *PerDeTemp* features another interesting characteristic: after having selected an optimal amount of resources for an application, it maps the allocation on the hardware by taking into account **Per**formance, aging and fault-induced performance **De**gradation, and **Temp**erature of the available cores.

4.6.1 The Tempura Scheduling Policy

The resource policy that targets the ES use cases is called "TEMPURA." The name stands for Thermal and Energy Management through Power bUdget Resource

Allocation. The main idea underlying *TEMPURA* is sketched in Fig. 4.5. Basing on system temperature and battery discharge rate, *TEMPURA* computes a power budget for the system. The budget is then translated to resource constraints.

TEMPURA specifically targets outdoors, battery-based embedded systems, i.e., systems that, in order to stay online, must limit battery usage and try not to reach critical temperatures.

To better understand the overall picture, we can consider the low-end ES use case (more on that in the next chapters). In that scenario, the system is supplied by a solar panel and a battery. Whenever the solar panel cannot provide enough power to recharge the battery (e.g., in case of bad weather), the HARPA-OS must perform resource allocation in a battery-aware fashion, so that the systems stay up all over the day. To do that, the system monitor can specify to the HARPA-OS a target lifetime (e.g., "must stay on at least for other 5 h"). This time constraint, along with the current charge level of the battery, is used by *TEMPURA* to compute a power budget for the system.

On the other side, reaching critical CPU temperatures would trigger the Linux OS thermal safety action (system switch-off), which would also prevent the system from staying online. To do that, *TEMPURA* tries to correlate temperature with the system power consumption, and this results in a second power budget for the system.

Obviously, *TEMPURA* merges the battery- and temperature-related power budgets by selecting the strictest one. At this point, thanks to a power consumption characterization that is performed at design time on the target architecture, *TEMPURA* is able to translate the power budget into a resource budget. That is, the maximum amount of resources that can be allocated to application gets limited in order to force the system to meet the selected power budget.

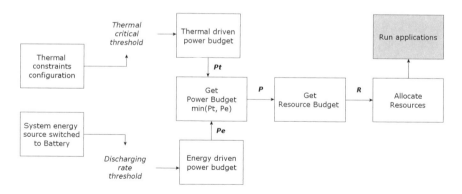

Fig. 4.5 The TEMPURA resource allocation policy block diagram

Fig. 4.6 ODROID XU-3 development board featuring the Samsung Exynos 5422 SoC

4.6.1.1 Design-Time Characterization

We carried out the experimental analysis on one of the HARPA project reference platforms, namely the ODROID XU-3 development board, which is based on the Samsung Exynos 5422 System-on-Chip. The interesting characteristic of this board is that it features an ARM big.LITTLE 8-core CPU (Fig. 4.6).

The processor is made by a high-performance cluster (4 Cortex A15 cores at 2 GHz maximum frequency), and a low-power one (4 Cortex A7 cores at 1.4 GHz max frequency), and it works in a heterogeneous multiprocessing (HMP) configuration. This enables the simultaneous execution of threads on both clusters at the same time. We sampled power consumption information by using the on-chip sensors, which reported the power consumption of the big (4 Cortex A15) and the LITTLE (4 Cortex A7) CPU clusters, the embedded GPU, and the memory. Furthermore, an external power meter with a sampling rate of 10 Hz allowed us to monitor the overall power consumption.

We stressed the system by executing a subset of multi-threaded OpenMP applications from the RODINIA benchmark suite, version 3.0 [3]: heartwall, hotspot, leukocyte, lavaMD, and srad.

Figure 4.7 shows the overall system power consumption and the contribution of each computing resource during the execution of the "heartwall" benchmark, which is a CPU-bound application. The total power consumption of the board fluctuates in the 12–14 W range, the average consumption observed for the big cluster is around 5 W, while the LITTLE cluster stays far below the 1 W.

As expected, the Cortex A15 quad-core, which is the most performing cluster, dominates the Cortex A7 quad-core in terms of power consumption. Therefore, it represents the most critical resource from the thermal management perspective. This is confirmed by the temperature measures shown in Fig. 4.8. We repeated the execution of the benchmark by comparing the case of CPU frequency governor set to "performance" (all the cores set to maximum frequency) to the case of governor

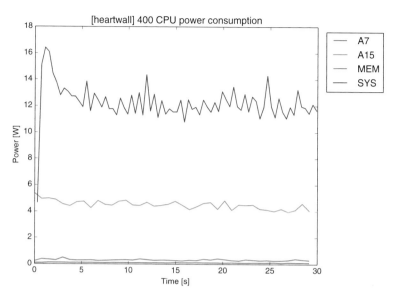

Fig. 4.7 Power consumption contributions during the execution of the "heartwall" RODINIA benchmark, running eight threads

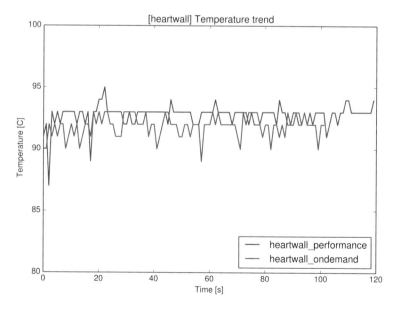

Fig. 4.8 ARM Cortex A15 quad-core ("big" cluster) temperature during the execution of the "heartwall" RODINIA benchmark, running eight threads

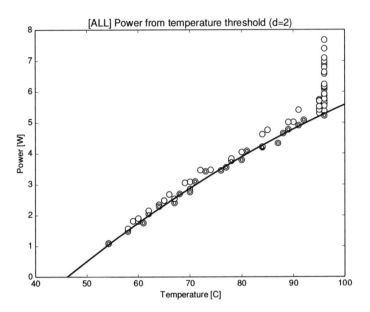

Fig. 4.9 Power consumption and temperature sampling of the ARM Cortex A15 big cluster. Double-circled dots belong to the Pareto frontier, which we also plotted (black line)

set to "on-demand" (frequency scaled up or down according to the load). In both cases, the processor temperature easily goes over 90 °C, reaching peaks of 96 °C after few seconds, despite the presence of a heat sink and a fan on top of the chip.

We performed a characterization of the ODROID XU-3 platform. In particular, we studied the power and thermal profile of the Cortex A15 "big" cluster, and the relationship between the total power consumption of the board and the sum of the power consumption contributions of each computing resource.

The scatterplot from Fig. 4.9 reports the temperature and power consumption of the big CPU cluster while running each benchmark for 1 min with a number of threads big enough to fully load all the available CPU cores.

Any given power consumption level can be translated to a small temperature range. However, since we aimed at guaranteeing a thermal safety threshold, we focused on the points that belonged to the Pareto frontier (such points are highlighted with double circles). We computed the analytic model of the frontier by performing a regression test. The result, which is also plotted in the figure, is shown in Eq. (4.1).

$$P_{A15} = -0.005 \times T_{A15}^2 + 0.187 \times T_{A15}^2 - 7.413 \qquad (4.1)$$

Basically, we bound the power consumption of the Cortex A15 quad-core cluster (P_{A15}) to the temperature value of the Pareto frontier (T_{A15}). Then, we investigated the relationship between the power consumption and the CPU usage of the big

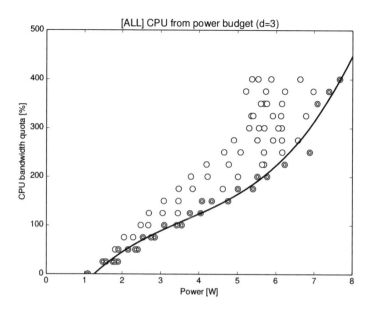

Fig. 4.10 CPU bandwidth allocation and power consumption sampling for the ARM Cortex A15 big cluster. Double-circled dots belong to the Pareto frontier, which we also plotted (black line)

cluster. To this purpose, we repeated the same regression analysis on the data set plotted in Fig. 4.10, where the power consumption of the CPU cluster has been put in relationship with the amount allocated CPU time (expressed in terms of CPU bandwidth: 100% means a whole core at full usage).

The resulting model is presented in Eq. (4.2).

$$C_{A15} = 2.20 \times P_{A15}^3 - 23.81 \times P_{A15}^2 + 120.31 \times P_{A15} - 117.62 \qquad (4.2)$$

Given a power budget P_{A15} for the big CPU cluster, we can translate it to an upper bound for the allocable CPU bandwidth quota. Unfortunately, as evident from the picture, this model does not accurately fit with the data set. In particular, the model results to be too conservative. Therefore, we considered the possibility to extract a specific model for each application. This is a feasible approach: applications that are managed by the HARPA-OS are known and are usually profiled at design time. In particular, each application can list its model in its recipe file. We therefore used the model from Eq. (4.3).

$$C_{A15} = p_3 \times P_{A15}^3 + p_2 \times P_{A15}^2 + p_1 \times P_{A15} + p_0 \qquad (4.3)$$

The parameters are specific for each benchmark. They are listed in Table 4.1.

Let us now focus on the power consumption of the whole system. The on-chip power consumption monitors allow us to retrieve the power consumption of the

Table 4.1 CPU usage—power consumption: application-specific model parameters (see Eq. (4.3))

Benchmark	p_3	p_2	p_1	p_0
heartwall	2.3	−23.3	110.3	−104.0
hotspot	3.6	−34.2	170.2	−158.5
lavaMD	0.9	−8.7	75.9	−78.1
leukocyte	0.4	−2.2	49.0	−53.9
srad	1.9	−18.2	115.1	−112.8

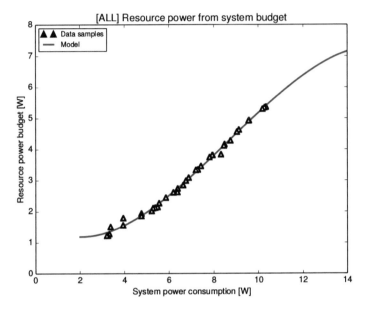

Fig. 4.11 System power consumption sampling and related sum of power consumption contribution of each resource

processor and the memory. However, in order to drive the battery lifetime, we must be able to compute the total board power consumption. In the following set of tests, we will therefore try to understand the relationship between power consumption of processor and memory and overall board power consumption.

Figure 4.11 shows the total system power consumption measured with an external power meter during the execution of each of the aforementioned benchmark applications, along with the sum of the power consumption contributions coming from the big cluster (4 x A15), the LITTLE cluster (4 x A7), and the memory. As the reader can guess by viewing the picture, this dataset can be accurately modeled. The result is shown in Eq. (4.4), where P_{res} is the power consumption of the computing resources (memory and CPU cores), and P_{sys} is the power consumption of the whole board.

$$P_{res} = -0.005 \times P_{res}^3 + 0.137 \times P_{res}^2 + 0.49 P_{res} + 1.656 \quad (4.4)$$

4.6.1.2 The Scheduling Algorithm

The pseudo-code of the policy is shown in Algorithm 1. First of all, the policy configures the number of allocation slots, which are a way to allocate to applications an amount of resources that is proportional to their priority (lines 1–4). Then, the policy computes which is the maximum power budget that allows the active applications to run on the available computing resources without causing the system temperature to reach the user-defined threshold (lines 5–20). As already stated, the temperature-related power budget is computed by using a thermal model that must be extracted from each application during an off-line phase. If an application does not provide it, the policy exploits the global model as defined in Eq. (4.1). If the system is running on batteries, the policy also computes the maximum power budget

---

Algorithm 1 Pseudo-code of the TEMPURA resource allocation policy
1: $numSlots \leftarrow 0$
2: **for all** application $\in$ ActiveApplications **do**
3:     $numSlots \leftarrow numSlots + application.Priority()$
4: **end for**
5: $boundResource \leftarrow$ NULL
6: $powerBoundMax \leftarrow 0$
7: **for all** resource $\in$ ManagedResources **do**
8:     $powerBound[resource] \leftarrow$ MAXVALUE
9:     **for all** application $\in$ ActiveApplications **do**
10:         $power \leftarrow GetPowerFromTemp(application,resource,TEMP\_THRS)$
11:         **if** $power < powerBound[resource]$ **then**
12:             $powerBound[resource] \leftarrow power$
13:             $boundApp \leftarrow$ application
14:         **end if**
15:     **end for**
16:     **if** $powerBound[resource] \geq powerBoundMax$ **then**
17:         $boundResource \leftarrow$ resource
18:         $powerBoundMax \leftarrow powerBound[resource]$
19:     **end if**
20: **end for**
21: $powerBudgetFromT \leftarrow powerBound[boundResource]$
22: $powerBudgetFromE \leftarrow GetPowerFromEnergy(TIME)$
23: $powerBudget \leftarrow \min(powerBudgetFromT, powerBudgetFromE)$
24: $budget \leftarrow \emptyset$
25: **for all** resource $\in$ ManagedResources **do**
26:     $budget[resource] \leftarrow GetResourceBudget(resource,powerBudget,boundApp)$
27: **end for**
28: **for all** application $\in$ ActiveApplications **do**
29:     $allocated \leftarrow \emptyset$
30:     **for all** resource $\in$ ManagedResources **do**
31:         $resourceSlotSize \leftarrow budget[resource]/numSlots$
32:         $allocated[resource] \leftarrow resourceSlotSize * application.Priority()$
33:     **end for**
34:     $Schedule(application,allocated)$
35: **end for**

that would allow the system to stay on until the user-defined time instant (line 22). At this point, the policy computes the system power budget as the strictest of the two aforementioned budgets (line 23).

Once the power budget is known, the policy translates it to a resource budget for each managed resource (lines 24–27). The LITTLE cluster is entirely allocated as a shared resource, as well as the system memory, since their contribution on the system power consumption is much lower than the big cluster contribution. Finally, the policy iterates over the active applications and computes the amount of resources to reserve them (lines 28–35).

4.6.1.3 Thermal Management Experimental Results

To validate the TEMPURA policy, we performed several tests, first focusing on the thermal management side, then on the energy side.

The first set of tests aimed at demonstrating that the TEMPURA policy could guarantee that the processor temperature would not surpass a user-defined threshold. We executed each benchmark separately under a thermal threshold of 75, 80, 85, and 90 °C. We stopped their execution after 40 s, since it is long enough to raise the CPU temperature up to the steady state [12].

The results are shown in Figs. 4.12 and 4.13. The efficacy of the policy is quite evident, since the temperature of the big CPU cluster remains below the specified threshold during the entire execution of the benchmarks. In the most critical scenario (threshold = 90 °C), the temperature goes slightly above the threshold for 3 s, but it stays otherwise under the threshold.

We repeated this set of tests in a multi-application scenario. More in detail, we co-run couples of benchmarks under the aforementioned temperature thresholds. Even this time, temperature results to be lower than the threshold most of the time.

Overall, *TEMPURA* is very effective at constraining the system temperature. Indeed, in most cases, *TEMPURA* results to be exceedingly conservative; however, it is worth remarking that these results have been obtained by using simple models and in a completely proactive fashion. This approach does not require to perform a continuous online monitoring of the temperature.

4.6.1.4 Energy Management Experimental Results

We executed a second series of experiments to assess the effectiveness of TEM-PURA in guaranteeing a required system uptime, i.e., a given system power budget.

We considered a realistic scenario: the system was supplied by a Li-on battery (3500 mAh, 3.70 V). We tested three lifetime goals: 2.5, 2, and 1.5 h, from which derives a system power budget of 5.18 W, 6.48 W, and 8.63 W, respectively. The results are shown in Figs. 4.14 and 4.15. Each plot reports the system power budget that we aim to fulfill in order to guarantee the system lifetime goal (dotted line), and the observed system power consumption (continuous line). The filled area represents

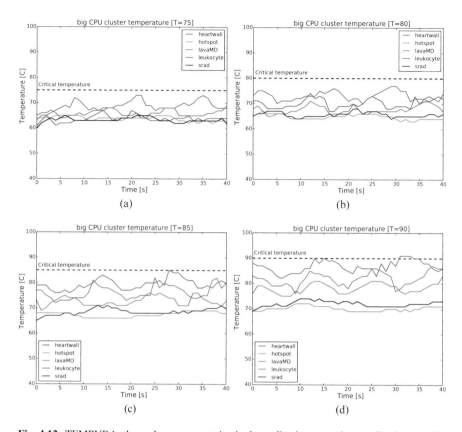

Fig. 4.12 TEMPURA: thermal management in single application scenarios (application-specific characterizations). (**a**) Threshold: 75 °C. (**b**) Threshold: 80 °C. (**c**) Threshold: 85 °C. (**d**) Threshold: 90 °C

the energy consumption, and the labels in overlay summarize energy budget and energy consumption, considering an execution time window of 30 s. Despite the presence of a few small system power budget violations, the results show that the TEMPURA approach is quite effective in guaranteeing a system lifetime goal. We can observe that in all the testing scenarios, the consumed energy is less than (but close to) the overall energy budget. In fact, the amount of estimated remaining energy is included in the 9–18% range, which is already a very good result, considering that the energy management strategy is performed in a completely proactive fashion.

Finally, we repeated the tests in a multi-application scenario. We executed couples of benchmarks under a system life expectancy of 1.5, 2.0, and 2.5 h. The results are summarized by Fig. 4.16, which shows the percentage of energy budget consumed for each scenario. Even in this case, even if TEMPURA results to be exceedingly conservative in some cases, the power budget is never exceeded.

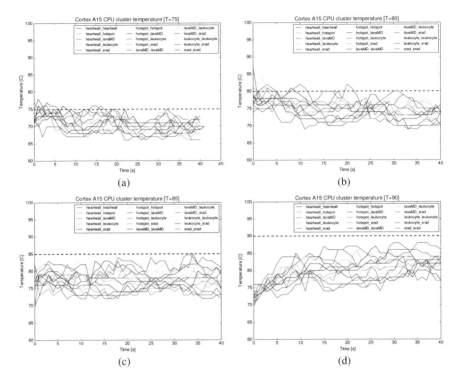

Fig. 4.13 TEMPURA: thermal management in multi-application scenarios (application-specific characterizations). (**a**) Threshold: 75 °C. (**b**) Threshold: 80 °C. (**c**) Threshold: 85 °C. (**d**) Threshold: 90 °C

4.6.2 The PerDeTemp Scheduling Policy

As already mentioned, HARPA-OS employs pluggable schedule policies. For the HPC scenario, the employed policies are two: YaMS, which is the HARPA-OS default scheduling policy and was used during the early stages of the project; and PerDeTemp, which is a new scheduling policy that is suitable for both HPC and homogeneous embedded scenarios and was used in the later stages of the project.

The difference between the two policies consists in how they compute the amount of resources to be allocated to applications. YaMS allocates discrete amounts of resources based on a limited set of known "golden" configurations. This means that, at any given time, the chosen allocation is one of those profiled at design time and inserted in the application recipe. This approach has two main disadvantages: first of all, it does not take into account the runtime status of the system, i.e., number of running applications, contention on shared resources, performance variability due to aging and faults, and the effect of different datasets on the behavior of applications; second, depending on the performance goal of applications, which

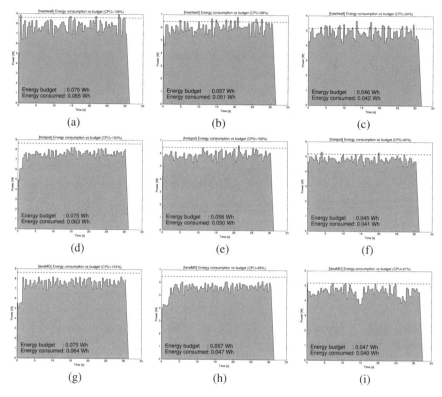

Fig. 4.14 TEMPURA: energy management in single-application scenarios. We computed the power budget in order to let the system stay up for 1.5, 2.0, and 2.5 h. (**a**) heartwall (1.5 h). (**b**) heartwall (2.0 h). (**c**) heartwall (2.5 h). (**d**) hotspot (2.5 h). (**e**) hotspot (2.5 h). (**f**) hotspot (2.5 h). (**g**) lavaMD (2.5 h). (**h**) lavaMD (2.5 h). (**i**) lavaMD (2.5 h)

is runtime-variable, the profiled configurations may cease to be "golden," and this often forces applications to run in sub-optimal configurations.

Conversely, PerDeTemp employs a continuous allocation: instead of relying on the applications' design-time profiling; it *entirely* bases allocation choices on the applications' runtime behavior. Each application, in fact, dynamically sends runtime feedbacks to the HARPA-OS in order to express its satisfaction with the current allocation, and PerDeTemp uses this information to compute which is the ideal amount of resources that must be allocated to applications. Another interesting feature of PerDeTemp, whose name stands for "**Per**formance, **De**gradation and **Temp**erature," is that resource mapping takes into account performance, faults-induced degradation, and temperature of the CPU cores. In particular, PerDeTemp strives in minimizing the resource consumption of applications while evenly spreading heat throughout the chip and mitigating the effects of performance degradation.

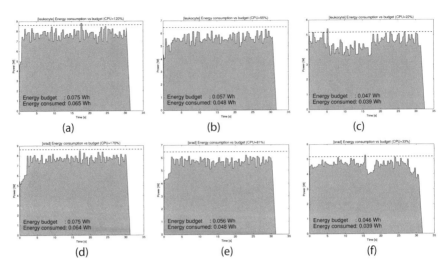

Fig. 4.15 TEMPURA: energy management in single-application scenarios (cont.). (**a**–**c**) leuko-cyte (2.5 h). (**d**–**f**) srad (2.5 h)

Both policies exploit the concept of Goal Gap, which is the percent distance between the current and the required throughput applications. The application notifies this information to the HARPA-OS at runtime, making it capable of reacting to allocation inefficiencies.

4.6.2.1 YaMS: A Multi-Objective Scheduling Policy

YaMS is a resource allocation policy that employs a multi-objective optimization algorithm and runtime feedbacks to dynamically compute which AWM is the best suited to comply with the declared performance goal. As already said, YaMS performs a discrete allocation: it chooses the best allocation out of the existing AWMs. Each AWM represents a "golden configuration" for the application, that is, a configuration that complies with a given QoS level that, in HPC scenarios, is usually the desired throughput; thus, the application recipe (the collection of AWMs) has to be carefully crafted using off-line design space exploration techniques. Figure 4.17 shows throughput and number of threads of Bodytrack application from PARSEC benchmark [1] running on a 16-core NUMA architecture. We express throughput in terms of *execution Cycles Per Second* (CPS), where "execution cycle" refers to the processing of a single chunk of data, i.e., an application Run phase (see Sect. 4.3). We ran Bodytrack with three different throughput goals: 1, 2, and 3 CPS. We intentionally removed the AWMs that complied with the 2-CPS and 3-CPS goals to show that YaMS is poorly able to deal with goals that have not been profiled during the off-line phase. The 1-CPS goal is met without problems because one of the AWMs contained in the recipe exactly complied with the goal. The other

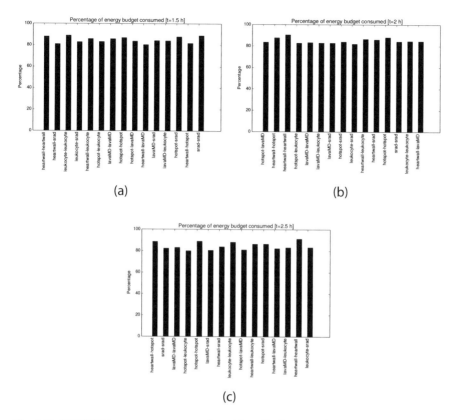

Fig. 4.16 TEMPURA: energy management in multi-application scenarios. (**a**) lifetime = 1.5 h. (**b**) lifetime = 2.5 h. (**c**) lifetime = 2.5 h

goals are only met on average, since there are not AWMs that comply with the goal; therefore, the policy continuously swaps between "too fast" and "too slow" configurations, thus becoming unstable. Note that this scenario is quite common: profiling information could be spoiled by the presence of co-runners (although that is not so common in HPC scenarios), frequency variations, and faults.

Early during the project, we performed some tests to evaluate the benefits that can be achieved by taking into account the runtime behavior of applications while computing resource allocation. Given that this approach resulted to be very effective, we extended the YaMS scheduling policy with the concept of "Goal Gap," which expresses the satisfaction of each application with its resource allocation. In YaMS, the Goal Gap information is used to understand whether an application is really reaching the performance value that has been measured during the off-line application characterization. If the observed performance is lower than expected, the allocation is temporarily switched to the next (hence more performing) allocation, so that the average observed performance gets closer to the expected one.

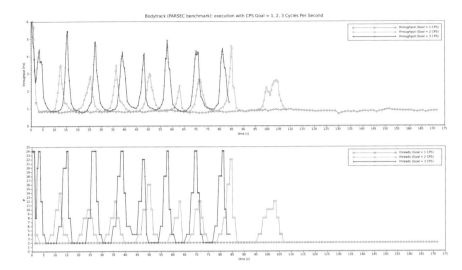

Fig. 4.17 Bodytrack application (from PARSEC benchmark) running on a 16-core NUMA architecture with a throughput goal equal to 1, 2, and 3 execution Cycles Per Second (CPS) (tolerance: 10%). We report the throughput [CPS] (upper figure) and the employed number of threads (lower figure). The oscillatory behavior is due to the fact that the application recipe does not contain AWMs that are specifically designed to achieve some of the throughput goals

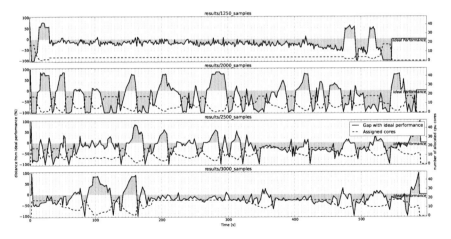

Fig. 4.18 Throughput (continuous line) and number of allocated CPU cores (dashed lines) of 4 Floreon+ Uncertainty instances running on a 48-core based system

Figure 4.18 shows the results of the execution of four instances of the Floreon+ Uncertainty application [8] on a 48-core NUMA system (8 six-core AMD Opteron 8425HE at 2.1 GHz). In this experimental scenario, each instance simulates the monitoring of a specific geographical area, receiving input from a dataset representing extremely variable forecast data and catchments levels, and carrying

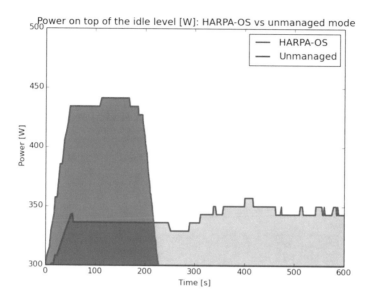

Fig. 4.19 System power consumption of 4 Floreon+ Uncertainty instances running on a 16-core multiprocessor NUMA system

out predictions about possible flooding. Depending on the input information, the application may need to increase or decrease the prediction rate, i.e., its throughput goal. This variability is managed by notifying the Goal Gap of each instance to the HARPA-OS. As previously discussed, HARPA-OS (through "YaMS") manages this information by balancing the resource allocation among the applications accordingly. The figure clearly shows the aforementioned instability effects: although the average throughput of applications is acceptable (all the instances are able to terminate before the deadline), the Goal Gap goes up and down depending on variability of the input data, and the "YaMS" capability of matching the applications' performance requirements with a suitable AWM.

The main goal of this experiment was to minimize power consumption by constraining resource usage of applications so that they terminated very close to—but never after—their deadline. The experiment resulted to be quite effective. Moreover, as shown in Fig. 4.19, the HARPA-OS and YaMS do not induce energy inefficiency: in fact, the energy consumption with and without resource manager are roughly characterized by the same energy consumption.

4.6.2.2 The PerDeTemp Scheduling Policy

Conversely to TEMPURA and YaMS, the PerDeTemp scheduling policy does not rely on a design-time characterization of application. On the contrary, it can manage even completely unknown applications by analyzing their behavior at runtime.

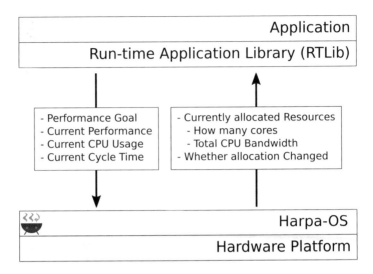

Fig. 4.20 Runtime profiling performed by the HARPA-OS Application Run-Time Library (RTLib)

As shown in Fig. 4.20, the HARPA-OS Application Run-time Library (RTLib) provides services to forward data from the application to the HARPA-OS daemon and vice versa. This data includes performance goal and runtime statistics from application side, amount of allocated resources from the HARPA-OS side. This process allows the HARPA-OS to understand whether (and to what extent) the application is satisfied with its current resource allocation.

Once started, applications can notify their performance goals to the HARPA-OS. A performance goal is again expressed in terms of CPS, where (as in YaMS) each cycle is the execution of a single Run phase. After each execution cycle, the RTLib updates a set of statistics that is called application runtime profile and contains the values of performance goal, current performance, cycle time, and CPU usage. The runtime profile is then forwarded to the HARPA-OS periodically or only when needed, depending on the application preferences. Upon forwarding a runtime profile, the RTLib computes the current Goal Gap, which is the percent distance between current and desired performance. The Goal Gap is computed as defined in Eq. (4.5), where $CPS_{average}$ and CPS_{ideal} are the observed (average) and required throughput, respectively.

$$\text{gap} = \frac{CPS_{average} - CPS_{ideal}}{CPS_{ideal}} \tag{4.5}$$

As shown in the equation, a negative Goal Gap means that the current performance is lower than the desired one and, conversely, a positive Goal Gap means that the current performance is higher than the desired one. The HARPA-OS clearly benefits from knowing the runtime profiles of applications; furthermore, the communication between applications and the HARPA-OS is bidirectional: each time the HARPA-OS changes a resource allocation, it also sends a detailed description of the new allocation to the involved application. Knowing the amount of resources they have at their disposal allows applications to reconfigure their software parameters and thus to suitably adapt to the new allocation. For instance, an application can dynamically change the number of running threads according to the available number of cores. Each time the HARPA-OS notifies an application about a new allocation, the application automatically enters into a Configure phase; during this phase, the application can use the RTLib API to query the amount of allocated resources and reconfigure accordingly.

The PerDeTemp policy employs learning techniques to dynamically compute the ideal amount of CPU bandwidth to allocate to each application. The entire process starts from the applications' runtime profiles. Upon receiving a runtime profile, the policy computes the amount of resources that are really needed by the application. Such amount is computed as shown in Eq. (4.6). Please note that this equation assumes a linear relationship between CPS and CPU quota allocation. This is not strictly true, hence the presence of learning techniques to adjust the allocation according to current hardware and workload.

$$\text{IdealBandwith}_{\text{CPU}} = \frac{\text{CurrentBandwidth}_{\text{CPU}}}{1 + \text{Gap}} \qquad (4.6)$$

As shown in Fig. 4.21, PerDeTemp allocates resources to applications trying to achieve an optimal performance, i.e., a Goal Gap equal to zero. This is done by learning at runtime which are the maximum and minimum allocations that cause the performance of the application to stick with the desired one. Such allocations are called boundaries, and each boundary is tagged with a resource amount, a goal gap, and an age. In the example, the boundaries are initially unknown. At second 1, the application forwards its runtime profile to the HARPA-OS, stating a Goal Gap of $1, 5$ for an actual CPU bandwidth of $4, 97$ CPUs. The resulting boundary is tagged as $<\text{ConstraintType, bandwidth, gap, age}> = <'\text{UpperBound}', 4.97, 1.5, 0>$, where age $= 0$ means that the boundary has just been created. The lower bound, conversely, is inferred from the upper bound and is equal to $<\text{ConstraintType, bandwidth, gap, age}> = <'\text{LowerBound}', 1.99, 0.0, -1>$. Each time a boundary is updated, the other one gets older. Once the age of a boundary reaches a customizable threshold, it is tagged as outdated and discarded.

Regarding resource mapping, as already mentioned, the set of cores that is allocated to each application is selected so that the cores feature a minimum or at least homogeneous degradation. During the process of selection, cool cores are preferred to hot ones, and the resource allocation is periodically recomputed to level the heat over the whole chip. To allocate resources in a performance-, degradation-, and temperature-aware fashion, PerDeTemp needs information about the status of both applications (current performance and current resource usage) and hardware (current degradation and temperature of each core). As shown in Fig. 4.22, this

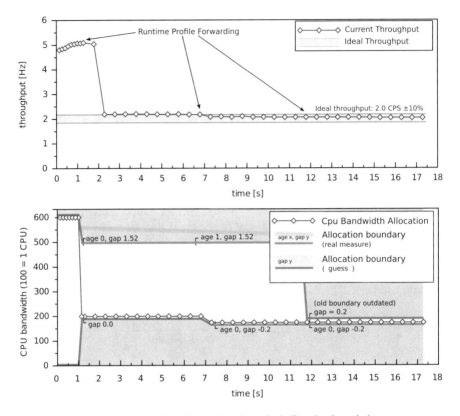

Fig. 4.21 PerDeTemp resource allocation policy: dynamical allocation boundaries

is possible due to the HARPA-OS, which gathers this information from multiple sources:

- The runtime library that is linked by applications automatically notifies the BarbequeRTRM about applications' statistics such as current CPU usage and satisfaction with the current CPU bandwidth budget;
- The fault prediction framework notifies the BarbequeRTRM about the presence and entity of degradation for each CPU core;
- The on-chip sensors (if available) are used by the BarbequeRTRM to periodically retrieve temperature values for each core.

The resulting thermal and degradation maps are used to compute an ideal resource mapping for the running applications.

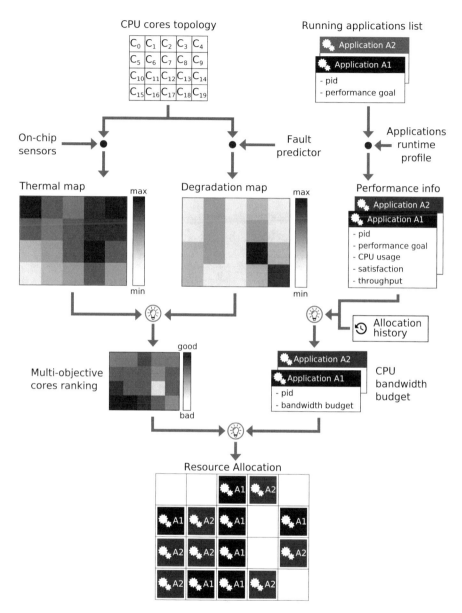

Fig. 4.22 The PerDeTemp allocation policy. The CPU cores' topology is annotated with the information coming from temperature sensors and fault prediction modules. This results in a multi-objective cores' ranking that will be used to compute the resource mapping. CPU bandwidth budget is instead computed by exploiting the information that comes with the runtime profiles. In particular, runtime profiles contain a satisfaction metric that is used by the runtime library to express whether the current CPU budget is under- or over-provisioned. Resource allocation is the union of bandwidth allocation (i.e., how many resources) and CPU mapping (i.e., which resources)

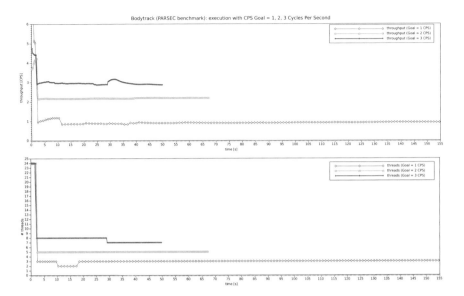

Fig. 4.23 Bodytrack application (from PARSEC benchmark) running on a 16-core NUMA architecture with CPS goal equal to 1, 2, and 3 CPS (tolerance: 10%, runtime profile sent every five cycles)

4.6.2.3 Experimental Results

Whereas we will validate the PerDeTemp resource mapping policy when presenting the HPC use cases, in this subsection we will validate the policy from the applications' throughput side.

AWM-less resource allocation policies converge to optimal solutions more easily than recipe-based ones. The reason is simple: in the case of AWM-less policies, resource allocation is continuous (i.e., the policy can allocate any amount of resources to applications); conversely, in the case of recipe-based policies, the allocation is discrete (i.e., the policy allocates resources according to existing configurations).

Figure 4.23 shows throughput and number of threads of the Bodytrack application under three different performance goals: 1, 2, and 3 CPS. Conversely to YaMS, which continuously swaps between configurations to reach an average null Goal Gap, PerDeTemp computes configurations on the fly, trying to continuously stick with the goal. Even if applications' reconfiguration (in this case, number of threads reconfiguration) introduces some noise on the cycle time, PerDeTemp succeeds in making the application comply with its declared throughput goal.

Another example of performance-aware resource allocation is shown in Fig. 4.24, where the policy is tested against a dynamic CPS goal. This scenario is realistic: an application could react to external events by updating its desired performance at runtime, e.g., to put itself in a low-power mode or to temporarily augment the quality of its output when needed. Even in this case, PerDeTemp succeeds in making the application comply with its goals even if, particularly in this case, application

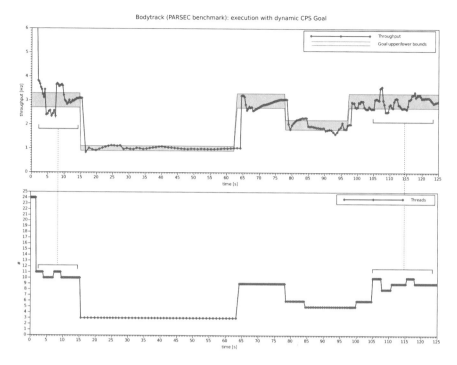

Fig. 4.24 Bodytrack application (from PARSEC benchmark) running on a 16-core NUMA architecture with dynamic CPS goal (tolerance: 10%, runtime profile sent every five cycles)

reconfiguration introduces some noise on the measured throughput (phases marked in red). Such a phase happens at second five, where the application runs ten threads—resulting in a negative Goal Gap—and then eleven threads—resulting in a positive Goal Gap. Indeed, this problem is quite similar to the one of relying on discrete allocations instead of continuous ones. This time, however, the problem cannot be solved, since the number of threads is inherently discrete. Having learned the behavior of the application, PerDeTemp finally allocates to the application an amount of resources such that the application is able to stick with the desired throughput while running ten threads.

It is worth to notice that the reaction time of the policy depends on how often runtime profiles are sent to the HARPA-OS (in this case, five execution cycles). Higher forward frequencies result in faster reaction times but also in higher communication frequency (and thus synchronization overheads). There is not an ideal value for the window size: it actually depends on the desired CPS. For example, in the case of a video encoder, the typical goal could be 25 frames per second, which is 25 CPS if the encoder processes a single frame per execution cycle. In this case, the forward frequency could easily be quite higher than that of an application like Bodytrack, whose goal is 1–3 CPS. Please note that the HARPA-OS defines a standard forward frequency, but each application can explicitly declare its own one.

4.7   Conclusions

In this chapter, we introduced the structure and the main features of the HARPA-OS. This framework, which acts at software level, employs task scheduling—hence the mapping of computing resources—as a software knob to tackle faults, aging, and performance variability.

After having introduced the concept of recipe, i.e., a set of golden resource allocation configurations for an application, we explained the motivations that drove us to move from a pure off-line profile-based task scheduling to more dynamical approaches.

We introduced two HARPA-OS scheduling policies: TEMPURA, which targets the Embedded Systems use case; and PerDeTemp, which was used both in the Embedded Systems (in a stripped version) and in the High-Performance Computing use cases (more will follow in the next chapters).

TEMPURA, whose allocation choices are partially based on an off-line profile of applications, tries to maximize performance while: (a) keeping the temperature of the processor below a critical threshold; and (b) guaranteeing the target system uptime by complying with an energy budget.

PerDeTemp, which is instead completely recipe-less and is hence able to handle completely unknown applications, aims at guaranteeing that applications comply with their expected QoS level, while: (a) counteracting performance variability; and (b) minimizing and evenly spreading temperature.

References

1. Bienia, C., Kumar, S., Singh, J., & Li, K. (2008) The parsec benchmark suite: Characterization and architectural implications. In *Proceedings of the 17th International Conference on Parallel Architectures and Compilation Techniques* (pp. 72–81).   New York: ACM.
2. Bolchini, C., Carminati, M., Gribaudo, M., & Miele, A. (2015) Task scheduling strategies to mitigate hardware variability in embedded shared memory clusters. In *DAC*.
3. Che, S., Boyer, M., Meng, J., Tarjan, D., Sheaffer, J. W., Lee, S.-H., et al. (2009). Rodinia: A benchmark suite for heterogeneous computing. In *IISWC 2009. IEEE International Symposium on Workload Characterization, 2009* (pp. 44–54).
4. Chou, C. L., & Marculescu, R. (2011, March). Farm: Fault-aware resource management in NoC-based multiprocessor platforms. In *2011 Design, Automation Test in Europe* (pp. 1–6).
5. Haghbayan, M. H., Miele, A., Rahmani, A. M., Liljeberg, P., & Tenhunen, H. (2016, March). A lifetime-aware runtime mapping approach for many-core systems in the dark silicon era. In *2016 Design, Automation Test in Europe Conference Exhibition (DATE)* (pp. 854–857).
6. Hartman, A. S., & Thomas, D. E. (2012). Lifetime improvement through runtime wear-based task mapping. In *Proceedings of the Eighth IEEE/ACM/IFIP International Conference on Hardware/Software Codesign and System Synthesis. CODES+ISSS 2018* (pp. 13–22). New York: ACM. Available: http://doi.acm.org/10.1145/2380445.2380455
7. Huang, L., Yuan, F., & Xu, Q. (2009, April). Lifetime reliability-aware task allocation and scheduling for MPSoC platforms. In *2009 Design, Automation Test in Europe Conference Exhibition* (pp. 51–56).

8. Massari, G., Libutti, S., Portero, A., Vavrik, R., Kuchar, S., Vondrak, V., et al. (2015, August). Harnessing performance variability: A HPC-oriented application scenario. In *2015 Euromicro Conference on Digital System Design (DSD)* (pp. 111–116).
9. Mercati, P., Paterna, F., Bartolini, A., Benini, L., & Rosing, T. (2016). Warm: Workload-aware reliability management in Linux/Android. *IEEE Transactions on Computer-Aided Design of Integrated Circuits and Systems, PP*(99), 1.
10. Stathis, J. H., & Zafar, S. (2006). The negative bias temperature instability in MOS devices: A review. *Microelectronics Reliability, 46*(2), 270–286.
11. Sun, J., Lysecky, R., Shankar, K., Kodi, A., Louri, A., & Roveda, J. (2014). Workload assignment considering NBTI degradation in multicore systems. *Journal on Emerging Technologies in Computing Systems, 10*(1), 4:1–4:22. Available: http://doi.acm.org/10.1145/2539124.
12. Wang, Z., & Ranka, S. (2010). Thermal constrained workload distribution for maximizing throughput on multi-core processors. In *2010 International Green Computing Conference* (pp. 291–298).
13. Zhuravlev, S., Saez, J. C., Blagodurov, S., Fedorova, A., & Prieto, M. (2012). Survey of scheduling techniques for addressing shared resources in multicore processors. *ACM Computing Surveys (CSUR), 45*(1), 4.

Chapter 5
Event-Based Thermal Control for High Power Density Microprocessors

Federico Terraneo, Alberto Leva, and William Fornaciari

5.1 Introduction

The semiconductor industry is currently facing significant roadblocks to deliver improvements in microprocessor performance. One of these is efficient dissipation of the produced power. The failure of Dennard scaling [2] in deep nanometre architectures has resulted in an ever-worsening power density increase, but heat dissipation strategies have not kept the pace. This fact has led to the well-known dark silicon problem [3, 9], where power and thermal constraints limit the number of transistors that can be operated to an ever-decreasing fraction.

Research to overcome this limitation is divided in two complementary directions. One aims at increasing the power dissipation capability of integrated circuits through improved thermal design and heatsinking; the other one addresses the development of microarchitectural improvements and run-time policies that can increase the power efficiency.

In such a scenario, thermal management is a fundamental design challenge that plays a key role in counteracting the variability encountered in current hardware and software architectures. Thus, effective dynamic (or, in other words, run-time) thermal management solutions can push a many-core platform to its maximum performance subject to the constraint imposed by the need to remain within safe operating temperatures.

As part of the HARPA project, thermal management has been addressed both from a high-level perspective, through the TEMPURA policy at the HARPA-OS level, and at the firmware level of HARPA-RT, with the event-based thermal controller presented in this chapter. Thermal management at the HARPA-OS level is

F. Terraneo (✉) · A. Leva · W. Fornaciari
POLIMI, Milan, Italy
e-mail: federico.terraneo@polimi.it; alberto.leva@polimi.it; william.fornaciari@polimi.it

© Springer International Publishing AG, part of Springer Nature 2019
W. Fornaciari, D. Soudris (eds.), *Harnessing Performance Variability in Embedded and High-performance Many/Multi-core Platforms*,
https://doi.org/10.1007/978-3-319-91962-1_5

targeted at use cases where the platform is thermally intrinsically safe, for example because it is already endowed with a hardware thermal protection, and the only goal is to enhance reliability through an added coarse-grained management layer. Thermal management at the HARPA-RT level, conversely, is dedicated to perform fine-grained thermal control at the timescale required by the existing hardware and software variability.

5.2 A Brief Overview of the Thermal Issue in Microprocessors

Microprocessors and in general CMOS integrated circuits consume electrical power through several mechanisms.

First and foremost, switching power is dissipated when individual transistors close causing a current flow to charge the gate capacitance of the transistors to which they are connected. This power contribution depends on four main factors which are related by the well-known formula:

$$P = \frac{1}{2}\alpha C V_{dd}^2 f. \tag{5.1}$$

Of these four terms, the load capacitances C are subject to manufacturing process variability, but most importantly the switching activity α is subject to large run-time variability due to the workload experienced by the microprocessor. For example, during a cache miss the functional units of a core may be mostly idle, thus causing an abrupt drop in the current consumption. The other two terms, voltage and frequency, provide instead a knob, the well-known DVFS through which it is possible to control the system behaviour.

Leakage power is instead caused by deep nanometre transistors deviating from the behaviour of ideal switches, and causing a significant current flow even in the open state. Leakage power depends on the physical conditions of the transistors, which include process variability and, remarkably, temperature. Thus, although leakage power does not directly depend on the workload being executed by the microprocessor, it depends indirectly from it through the temperature rise caused by the switching power.

Other mechanisms that cause power dissipation include resistive losses in the on-chip metal interconnection, which again depend on the current flowing due to the other dissipation mechanisms.

One important fact to notice is that the power consumption of microprocessors is not constant. Active power causes current consumption variations at the timescale of the individual clock edges and the switching activity is heavily influenced by factors including microarchitectural variability induced by cache misses, code patterns exercising the functional units in different ways (think floating point intensive vs.

control flow intensive code fragments), inter-core communication and, at a higher level, OS task scheduling and time-varying workload.

All these factors add up and contribute to induce significant power variations ranging from the nanoseconds to the seconds timescale. Although smoothed by the decoupling capacitors always present in a microprocessor design, power dissipation has significant frequency content in a very broad range, and as we will see, this impacts the design of thermal control policies.

5.2.1   The Thermal Dissipation Problem in a Nutshell

The electrical power consumed by integrated circuits is converted into thermal power, and has to be efficiently dissipated. From a physical perspective, most of the power consumed by an integrated circuit is dissipated in the active silicon layer, whose height is only a fraction of the total chip thickness, and the chip thickness is already a small dimension, usually in the range of a few hundred micrometres. This localised heating causes the two main challenges of thermal control: high power density and fast thermal dynamics.

The first issue is mainly of concern to the research lines trying to design better thermal dissipation solutions, while the second is what shapes the requirements of thermal control policies.

Using the well-known electrical equivalent of thermal circuits, it is possible to sketch a simplified thermal model for a generic microprocessor (Fig. 5.1). This model is too simplistic for simulation, but its abstract nature is well suited to aid the reader in gaining a quick understanding of the involved phenomena.

The power generated by the processor (here assumed spatially uniform for simplicity) is modelled as a current generator P_{gen}. This power is first *quickly* absorbed by the thermal capacitance of the silicon chip $C_{silicon}$, and dissipated through the thermal resistance from the chip to the heatsink $R_{silicon}$. This in turn *slowly* heats up the heatsink thermal capacitance C_{sink}, which finally exchanges heat convectively with the ambient.

The key aspect that can be gained from this model is in the two highlighted adjectives of the previous sentence. The thermal dissipation stack can be seen, as a first approximation, as a two-stage RC filter over the microprocessor power consumption. Since the microprocessor chip has such a small volume, its thermal

Fig. 5.1 A simplistic model of the heat dissipation stack for a microprocessor

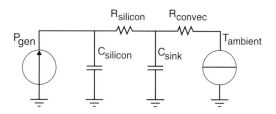

capacitance is quite low, and thus the chip temperature can change quite rapidly. The heatsink, on the other hand, is a large block of metal and its temperature changes much more slowly.

As we have seen before, the power of a microprocessor exhibits significant variations in a wide range of timescales, thus we are sure that, in real-world workloads, both thermal dynamics will be exercised, and since the objective of thermal control is that of controlling the chip temperature, not the heatsink one, control policies need to act at a faster timescale than the fastest involved dynamic to be of any effectiveness.

But, how fast can on-chip temperature vary? The best way to answer this question is through an experiment. To model a dynamic system, in the sense given to this term by control theory, it is possible to feed it with some reference signal and measure the output. Of the many possibilities, the step and impulse response are two of the most suitable ones. Their main advantage is the easy interpretation of the result as well as the possibility to reproduce the behaviour of the system when subject to arbitrarily complex signals by suitable convolutions. Thus, although real-world CPU workloads are far more complex, we have chosen a step response for the presented experiment.

The experiment was performed on an Intel Core-i5 6600K running Linux, and to produce the step in power consumption the program cpuburn [8] was used. The result is shown in Fig. 5.2.

The cpuburn program was started 5 s after temperature logging was enabled. In addition, it was briefly stopped for 500 ms at $t = 23$ s and $t = 41.5$ s to show an important point that will be discussed shortly.

Due to the small thermal capacitance of the silicon $C_{silicon}$ and the non-negligible thermal resistance towards the heatsink $R_{silicon}$, the processor temperature rises from 54 to 70 °C in 50 ms, a 16 °C increase, and reaches 80 °C after only 600 ms since cpuburn was started. After that, the heat produced by the processor starts heating up the heatsink, which in turn drags the chip temperature even higher, although more slowly. Since this second part of the thermal transient is driven by the heatsink thermal dynamics, it takes roughly 55 s, two orders of magnitude more time for the

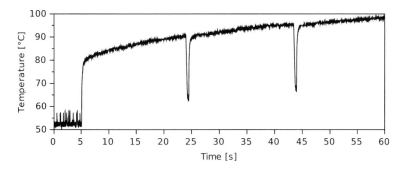

Fig. 5.2 Step response of the heat dissipation stack for an Intel processor

chip temperature to rise from 80 to 99 °C, when the experiment was stopped not to risk damaging the processor.

Summarising, this experiment evidences the existence of two well separate thermal dynamics. When the power is increased stepwise, the temperature will increase at a fast rate at first, and then continue increasing more slowly.

There is just one final remark to be made before considering the effect of these considerations on the design of thermal policies. By observing the behaviour of the system when cpuburn is first started, it can be noticed that although temperature rises at a fast rate, it does so when temperature is low, and when temperature is high, which is where thermal control policies should act, temperature increase is much slower. This is a common objection to the need for fast thermal control policies. Consider a hypothetical policy trying to keep the processor temperature below 85 °C. Since cpuburn is started at $t = 5$ s but the temperature first reaches 85 °C at $t = 10$ s, it appears that the policy has 5 s to react.

This is not true, as the fast temperature rise occurs at low temperatures *only when the heatsink is cold*. Consider again Fig. 5.2 and concentrate on what happens at $t = 23.5$ s. Here, cpuburn has been stopped, but since the heatsink takes a significant time also to cool down, when the power consumed by the processor is increased again, the core temperature reaches 85 °C in just 110 ms, and when the same operation is performed at $t = 42$ s when the heatsink temperature is even higher, only 54 ms are needed to reach 85 °C.

This clearly shows that thermal policies, to be effective, have to respond at most in a few tens of milliseconds.

5.2.2 Thermal Control Policies

We hope that by now we have convinced the reader that one of the most challenging tasks in designing a thermal control policy is that the policy sensing, decision and actuation loop needs to be run at the millisecond timescale due to the need to control the microprocessors' fast thermal dynamics.

A policy can be implemented entirely in hardware, with a dedicated data path and state machine implementing the policy algorithm, or can be implemented in software, where only the temperature sensor and DVFS actuator are hardware components. The main advantage of a hardware implementation is that it frees the microprocessor from the overhead of executing the thermal control algorithm, which given that it needs to be executed every few milliseconds can reduce the performance of applications noticeably. However, the main drawback of a hardware implementation is its inflexibility, as it prevents fine tuning that can be needed to overcome the software and hardware variability.

The proposed event-based thermal controller takes a mixed hardware–software approach to achieve the benefits of both a hardware and a software implementation. An ordinary PID control algorithm has to be executed at a fixed rate, even when temperature is not changing. In the HARPA project, we have explored the powerful framework of event-based control theory, to design a small hardware state machine which generates events only when temperature is changing. These events cause interrupts which in turn execute the software control algorithm. By doing so, the flexibility of a software implementation is preserved, while significantly reducing its overhead.

5.3 A Modelica Thermal Simulation Library

To perform the simulation studies required by the presented research, we used Modelica, which is an object-oriented modelling language.

Two are the main reasons for the choice of object-oriented modelling, and in particular the Modelica language: multi-physics modelling, and the possibility of co-simulating equation-based and algorithmic components. Multi-physics is important for the addressed domain, since achieving a sound validation of a thermal management policy requires to account for the many and diverse installation conditions that a microprocessor can encounter. It is thus necessary to model the chip, the heat sink, possibly the casing and the cooling system, be this just an air mover or a liquid cooler or anything else, and sometimes even the power supply electronics. The co-simulation of components described through equations and through algorithms allows to model the processor and its heat dissipation stack as well as the controller in their most natural representation, and also for the generation of *stimuli* to replicate the behaviour of the various computational loads to which the microprocessor can be subjected. Quite intuitively, a modelling paradigm that allows to do all of the above with a single tool is of great help for the presented research.

The thermal simulation and control has been implemented as a Modelica library, which is composed of three sections briefly described below:

1. *Components.* This section contains the building blocks used by the rest of the library, such as a solid control volume with heat capacity and boundary conductances, well-assessed empirical correlations for convective heat exchange and domain-specific elements like a cooling fan. The same section contains aggregate components, including a three-dimensional array of solid volumes, which is typically used to simulate the temperature distribution into a chip or a heat spreader. One-dimensional elements are also available, being suited to model elements such as heat pipes. The components of this section are directly modelled in the continuous time with differential equations. The separation of the modelling equations from the differential equation solver is in fact another important advantage of adopting an object-oriented paradigm, as this allows to take profit of newly developed solvers with no effort on the part of the analyst.
2. *Controllers.* The main element of this section is the event-based controller presented in Sect. 5.5; the section also contains the models of the aggregate blocks used to compose the devised control architectures, so as to obtain the results of Sect. 5.6. For the reasons explained above, all these models are

algorithmic. In addition, this section includes continuous-time controllers, that are functionally equivalent to the event-based ones just mentioned, and can be used as a baseline for comparison in simulations targeted to control quality assessment.

3. *Stimuli sources.* This section includes ad hoc power signal generators, useful to produce realistic profiles based on conveniently specified characteristics—such as the use over time of arithmetic units, pipelines, cache and so forth—of the possible applications running on the simulated microprocessor.

In developing the library, care was taken to design physical connectors (electric pins, heat ports and so on) compatible with the Modelica Standard Library, and with other libraries that can be usefully coupled to this in a view to widening the set of simulated physics—for example, the ThermoPower library [1] to represent thermal and hydraulic phenomena in cooling systems, including both those comprised in the typical CPU rack as well as for large-scale simulations of data centre air conditioners.

To briefly show the library in action, Fig. 5.3 shows the Modelica scheme of the model of a microprocessor, including the heat spreader which is part of the package, connected to a heat sink, exchanging with air at a prescribed ambient temperature, subjected to a time-varying (in this case, multi-harmonic) profile of generated powers in the absence of any thermal management.

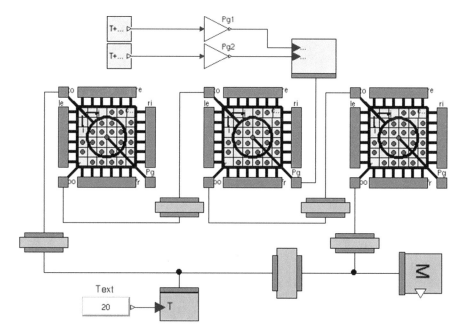

Fig. 5.3 Chip subjected to exogenous powers, open loop—Modelica scheme

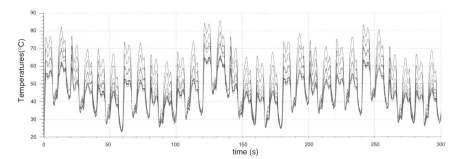

Fig. 5.4 Chip subjected to exogenous powers, open loop—ten simulated temperatures from the grid

Figure 5.4 shows an example of the obtained results, showing the behaviour of selected temperatures in the chip for clarity. Note the large and fast swings, furthermore testifying the criticality of the control problem and the usefulness of the event-based approach.

The simulation of Fig. 5.4, that was done with a 10×10 grid and three layers to represent the silicon, package and heat sink, took 23.5 s on a 64-bit i5-based machine. This is about 25 times faster than real time, allowing for extensive test campaigns with an acceptable effort, as well as giving the possibility to test corner cases in the silicon power consumption difficult to exercise using benchmarks on a real machine. As for the spatial discretisation, based on experience we can say that the adopted one is adequate for most control-oriented—i.e. system level—studies. At present, research is nonetheless ongoing towards the use of sparse solvers, in order to allow for significantly more fine-grained simulations, should this be necessary, e.g. for the final validation of a strategy. We would like to stress once again, however, that the tool is already well suitable for control-centred studies, as testified by the experimental results of Sect. 5.6, where the used control system has been assessed by means of the library just described.

5.4   Event-Based Thermal: The Hardware Event Generator

In [7], it is shown and motivated that a decentralised control architecture, devoting one event-based PI controller to each core, can provide efficient enough temperature control. In this and the following section, we describe how the preliminary results of that paper were turned into a fully functional system working on real hardware.

As anticipated in Sect. 5.2.2, hardware–software partition is the proposed solution to achieve the fast response required by the thermal control problem at an acceptable overhead. Specifically, and given the decentralised nature of the overall scheme, each core is equipped with a hardware event generator, to interact with the thermal sensors, and a software controller.

This section deals with the hardware event generator, the following with the software control policy.

The purpose of the hardware event generator is to generate interrupts to run the software control policy only when needed. The design combines a send-on-delta and a timeout policy as follows. The temperature sensor is sampled at a fixed interval q_s, typically in the range from 500 μs to 10 ms. A temperature threshold Δ and a timeout are selected. Every q_s, a new temperature value is read, and if it differs in magnitude from the value when the controller was last run by more than Δ, a threshold event is generated and the controller is run again. If instead the timeout expires, a timeout event is generated and the controller is run. The software controller is informed whether the event is due to a threshold or timeout through a flag bit in a register exposed by the event generator.

The timeout value is dynamically adapted (in software) to satisfy the opposing constraints of control quality and low overhead. The decision to increase or decrease the timeout depends on the reason why the controller was called. In the case of a timeout event, the timeout is increased exponentially, up to a maximum value that in the proposed implementation is 0.5 s. If instead the controller was called due to a threshold event, the timeout is immediately reduced down to the minimum value q_s, forcing the controller to be run again when the next temperature sample is available.

The event generation policy is expected to be implemented in hardware, as a simple state machine connected to a data path to compare the absolute value of the current temperature reading with the one when the controller was last run, hence deciding if an interrupt has to be generated. The timeout can be easily implemented using a hardware counter incremented at a frequency equal to $1/q_s$. Figure 5.5 shows a high-level diagram of the proposed event generator, detailing the registers used for the required communication with the software interrupt routine.

To have an estimate of the area and power consumption of the proposed hardware event generator, we implemented and simulated it in RTL Verilog, and then synthesised it in Cadence Encounter using the NAND Gate Liberty standard cell library. Synthesis results were obtained considering an operating voltage of 1.1 V and a clock frequency of 667 MHz, which yielded a per-core area and power

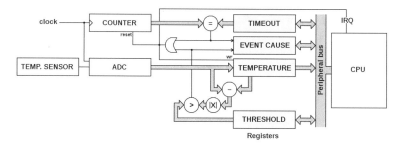

Fig. 5.5 High-level logic scheme of the hardware event generator

overheads of $159\,\mu\text{m}^2$, $471\,\mu\text{W}$, respectively. The proposed solution has thus a negligible impact compared to the area and power consumption of modern CPU cores.

5.5 Event-Based Thermal: The Software Control Policy

In this section we describe the software part of the proposed solution, that is, the configurable control policy.

5.5.1 Control Structure and Closed-Loop System

As discussed in [7], control events are generated by the temperature sensor with a send-on-delta policy, that now needs to be described more in detail. It is not the purpose of this section to describe the internals of the sensor, suffice in this respect to say that it has an internal "fast" sampling mechanism acting at period q, based on which an event is possibly triggered. We thus introduce an index h to count the mentioned fast sensor samplings.

As anticipated, we adopt a PI control law. In state space representation, this takes the form of two multiply-and-accumulate operations, that is,

$$\begin{cases} x_R(k) = x_R(k-1) + b_R(w(k-1) - y(k-1)) \\ u(k) = x_R(k) + d_R(w(k) - y(k)) \end{cases} \tag{5.2}$$

where w, y, u and x_R are the set point, the controlled variable, the control signal and the PI state variable, while k is the *control* discrete-time index—i.e. it counts the interventions of the controller, not the samplings of the sensor. To keep the notation as light as possible, in the following when we need to locate the k-th event in the fast sampling, we shall indicate its index with $h(k)$, while when this is not needed we shall write for the generic signal $v(k)$ to mean $v(h(k))$.

Based again on [7], we describe the controlled system with a first-order, SISO, strictly proper model. Also, we make the assumption that the fast sampling period q is small enough for a discrete-time model at step q to be practically equivalent to a continuous-time one. This said, we write

$$\begin{cases} x_P(h) = a_P x_P(h-1) + b_P u(h-1) \\ y(h) = c_P x_P(h) \end{cases} \tag{5.3}$$

where x_P is the state of the controlled system. The controller (5.2) is realised in event-based form as follows. At the k-th event, the sensor provides the current and also the previous sample of the controlled variable, i.e. $y(h(k))$ and $y(h(k)-1)$.

The controller then computes $u(k)$ and keeps it constant till the last event, that is,

$$u(l) = u(h(k-1)) = u(k-1), \quad l = h(k-1)\ldots h(k)-1. \tag{5.4}$$

In accordance with the hold operation above, prior to computing $u(k)$, the state x_R is made consistent with $y(h(k)-1)$. Assuming that the set point is modified—or sensed—only at events, which is sensible as it is seldom (if ever) modified, from the controller's viewpoint we have

$$w(l) = w(h(k-1)) = w(k-1), \quad l = h(k-1)\ldots h(k)-1. \tag{5.5}$$

Given the above, the controller state at time $h(k)-1$ is

$$\begin{aligned}
x_R(h(k)-1) &= u(h(k)-1) \\
&\quad -d_R(w(h(k)-1) - c_P x_P(h(k)-1)) \\
&= u(k-1) - d_R(w(k-1) - c_P x_P(h(k)-1)),
\end{aligned} \tag{5.6}$$

and therefore the same state at the k-th event is

$$\begin{aligned}
x_R(k) &= x_R(h(k)-1) + b_R(w(k-1) - c_P x_P(h(k)-1)) \\
&= u(k-1) - d_R(w(k-1) - c_P x_P(h(k)-1)) \\
&\quad +b_R(w(k-1) - c_P x_P(h(k)-1)) \\
&= u(k-1) + (b_R - d_R)(w(k-1) - c_P x_P(h(k)-1)).
\end{aligned} \tag{5.7}$$

Now, since

$$u(k-1) = x_R(k-1) + d_R(w(k-1) - c_P x_P(k-1)) \tag{5.8}$$

we get

$$\begin{aligned}
x_R(k) &= x_R(k-1) + d_R(w(k-1) - c_P x_P(k-1)) \\
&\quad +(b_R - d_R)w(k-1) \\
&\quad -(b_R - d_R)c_P x_P(h(k)-1) \\
&= \ldots \\
&= x_R(k-1) + b_R w(k-1) - d_R c_P x_P(k-1) \\
&\quad +(b_R - d_R)c_P x_P(h(k)-1).
\end{aligned} \tag{5.9}$$

If we evidence the variation of x_P from the $(k-1)$-th event till "immediately" (i.e., q) before the k-th as

$$x_P(h(k)-1) = x_P(k-1) + \delta x_P(k-1), \tag{5.10}$$

where the index attributed to δx_P is $k-1$ as that quantity is known before the k-th event, we obtain

$$
\begin{aligned}
x_R(k) = {}& x_R(k-1) - b_R c_P x_P(k-1) + b_R w(k-1) \\
& + (d_R - b_R) c_P \delta x_P(k-1).
\end{aligned}
\tag{5.11}
$$

Reasoning in an analogous way for the state of the controlled system, we have

$$
\begin{aligned}
x_P(k) ={}& a_P x_P(h(k)-1) + b_P u(h(k)-1) \\
& a_P x_P(h(k)-1) + b_P u(k-1) \\
& a_P(x_P(k-1) + \delta x_P(k-1)) + b_P u(k-1) \\
={}& \ldots \\
={}& (a_P - b_P d_R c_P) x_P(k-1) + b_P x_R(k-1) \\
& + b_R w(k-1) + (d_R - b_R) c_P \delta x_P(k-1).
\end{aligned}
\tag{5.12}
$$

Thanks to the holding mechanism described above, and to the pre-update of the x_R in accordance with the (additional) past value of the controlled variable transmitted by the sensor, we can represent the closed-loop system in the k index—i.e. counting the events independently of their distance in the constant-rate sampling at step q—by writing

$$
\begin{cases}
x(k) = Ax(k-1) + bw(k-1) + f\delta x_P(k-1) \\
o(k) = Cx(k) + dw(k)
\end{cases}
\tag{5.13}
$$

where $x(k) := [x_P(k)\, x_R(k)]'$, $o(k) := [y(k)\, u(k)]'$ and

$$
A = \begin{bmatrix} a_P - b_P d_R c_P & b_P \\ -b_R c_P & 1 \end{bmatrix}, \quad b = \begin{bmatrix} b_P d_R \\ b_R \end{bmatrix},
$$

$$
f = \begin{bmatrix} a_P \\ (d_R - b_R) c_P \end{bmatrix} \quad C = \begin{bmatrix} c_P & 0 \\ -d_R c_P & 1 \end{bmatrix}, \quad d = \begin{bmatrix} 0 \\ d_R \end{bmatrix}
\tag{5.14}
$$

Finally, and again in accordance with [7], we endow the event triggering mechanism with a timeout, i.e. a time span after which the sensor performs a new transmission unconditionally. This is useful as a keep-alive measure, and in no sense impairs our results.

5.5.2 Tuning, Stability, Robustness and Performance

The synthesis of (5.2) is done in the discrete-time domain. Doing so, a natural way to tune the PI controller is to prescribe the eigenvalues of matrix A in (5.14). For example, setting

$$b_R = \frac{1 + e_1^2 - 2e_1 - e_2}{b_P c_P}, \quad d_R = \frac{1 - 2e_1 + a_P}{b_P c_P} \qquad (5.15)$$

makes those eigenvalues $e_1 \mp \sqrt{e_2}$. It is therefore straightforward to make the closed-loop system asymptotically stable, while having the integral action in the controller structurally guarantee zero steady-state error, which also implies complete asymptotic rejection of constant disturbances.

Given the simplicity of the controlled system model described above, evaluating its stability degree, hence its robustness, is straightforward. More difficult is an assessment of its performance, however: it is easy to compute how many k steps— i.e. events—the free motion of (5.13) needs to converge to zero within a given tolerance, but this gives no information about the *time* duration of that transient, since (5.13) disregards for the interval between two adjacent events. This aspect deserves a few more comments.

First, denoting by $\overline{N}_c$ the number of steps taken by the free motion of (5.13) to converge to zero, a worst-case estimation of the closed-loop settling time (an adequate performance indicator for our purposes) is readily obtained as $\overline{N}_c$ multiplied by the prescribed timeout. This is inherently very pessimistic, however, because if most of the events during a transient are timeouts, then either the PI is poorly tuned or the send-on-delta threshold is not adequate.

An inherently optimistic settling time estimate, conversely, is obtained by supposing that the number of h steps between every two subsequent events equals the send-on-delta threshold divided by the one-step variation of y taken at the h step corresponding to the first of the two, of course rounded to the nearest greater integer. The settling time estimate is then the sum of so approximated inter-event periods up to the $\overline{N}_c$-th.

We omit further details on this matter because research is still underway, and in any case the results obtained to date are satisfactory, generally outperforming state-of-the-art alternatives.

In practice, then, to calibrate the control system, one can obtain the required model by subjecting the processor to a load and a DVFS step; from the so-gathered response data, and given the fast sampling time, model (5.3) is parameterised straightforwardly with any of the numerous methods available in the literature.

5.5.3   Implementation and Simulation-Based Assessment

When the proposed solution is implemented in a multi-core or many-core processor, the sensor and hardware event generator are conceived to be implemented in hardware (although for the tests of Sect. 5.6 we had to emulate this). The event-based controller, conversely, is expected to be realised in software. As anticipated, events are triggered when the present temperature differs in magnitude more than Δ from that at the last event, or when a configurable timeout T has elapsed since that event. If an event is caused by a timeout, the software controller increases the value of T up to a prescribed maximum. If on the contrary the event is caused by a temperature variation, T is brought back down to a prescribed minimum. This is substantially the same event triggering mechanism of [7], thus we omit further details if not for noticing that the conjectures made therein are now confirmed on real hardware.

The used knob is the DVFS mechanism present in any modern processor. More precisely, the event-based controller produces as output a frequency value. This value is decided based on the current temperature, the state of the controller and the desired set point, which is set to a value slightly lower than the acceptable maximum temperature. Since the controller produces a frequency value, the setting of the corresponding voltage is left to the installed electronics.

Modern operating systems also have power-performance controllers which measure the system load and use a "governor" to act on DVFS to increase the frequency when the load is high and decrease it in the opposite case, to save energy and useless heating while keeping the system responsive.

The presented thermal management scheme can seamlessly integrate with the power/performance system via an override configuration: the lower frequency is selected and applied between the one requested (based on the load) by the power/performance governor, and the one computed by the temperature PI controller, the state of which is updated accordingly. This configuration naturally makes power/performance prevail when the temperature is well below the limit, and thermal control takes action and rules in the opposite case.

To give an example of the assessment activities carried out by means of the presented Modelica simulation library, we show a test for the event-based controller, completed with all the features just listed, subjected to a suddenly variable computational system load. Figure 5.6 shows the controlled core temperature and the maximum admissible one T_{thr}, while Fig. 5.7 reports the frequency output from the controller, normalised in the (0,1) range, which corresponds to the available frequency span. The set point given to the event-based PI is computed as the maximum admissible temperature minus twice the send-on-delta threshold—an empirical rule of thumb that proved successful in practice.

As can be seen, despite the large and abrupt load variations, the objective of operating at the maximum speed required by the load, while not violating (if not in a practically negligible manner, which is admissible) the temperature limit, is met. In fact, given the typically low resolution of the temperature sensors embedded in microprocessors, one could even use the maximum admissible temperature as set point directly, and the benefits of the proposed control for the device would still be obtained in full.

To weigh the overhead of the proposed control scheme, Fig. 5.8 reports the inter-event times along the test. As can be seen, the fast reaction time of the proposed controller has been obtained by using a fast controller activation interval only when needed, i.e. when rapid temperature changes occur. The inter-event period is instead increased up to 100 ms for a significant part of the simulation. Thus, the remarkable

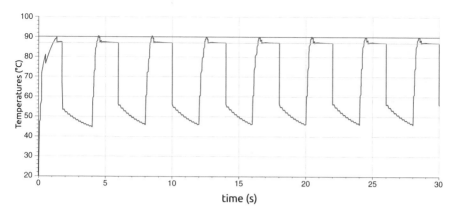

Fig. 5.6 Example of control assessment simulation—processor temperature (blue) versus the maximum admissible one (red) and the set point provided to the event-based controller (green)

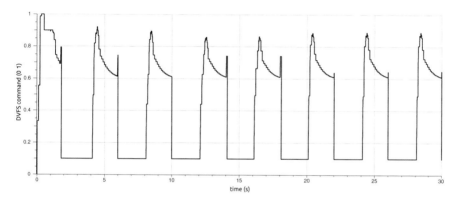

Fig. 5.7 Example of control assessment simulation—DVFS command normalised into the (0, 1) range

control performance of Fig. 5.6 is obtained with just 809 events in the 30 simulated seconds—i.e., about 27 events per second on average. This fact, confirmed in practice as in the performed experiments we could get even lower events/second ratios, indicates that also the objective of a low system overhead is achieved.

5.6 Experimental Results

The system we used for experimental tests is a PC with an ASUS Z170K motherboard equipped with an Intel i5-6600K microprocessor and a Cooler Master finned heat-pipe sink. The fan was removed from the heatsink during the tests, to

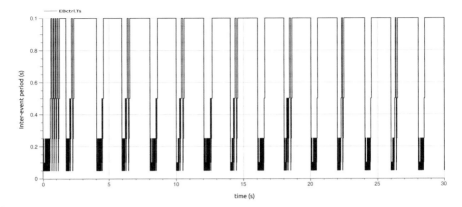

Fig. 5.8 Example of control assessment simulation—inter-event time

Fig. 5.9 Experimental setup

simulate an underdimensioned heatsink as well as to better assess the controller under extreme heat exchange conditions. A photo of the experimental setup is shown in Fig. 5.9.

The experiments were performed using Ubuntu Linux version 15.10. The kernel modules devoted to thermal and power/performance management were removed, and the userspace governor was selected, so as to avoid any operating system interference in the tests. Also, for a meaningful assessment, we disabled the Turbo Boost feature of the microprocessor, as it is based on internal hardware that is not required by our solution, and could unpredictably alter the results. As disabling Turbo Boost limits the maximum CPU frequency, we overclocked it to operate up to 4.2 GHz. The processor has internal thermal protections, named TM1 and TM2, that engage when the critical temperature (in the used processor, 100 °C) is reached. These protections were left in place, as they have no role if not in a thermal emergency, that however with the proposed controller never happened. Under this configuration, the processor DVFS is not controlled by any policy other than the proposed controller, thus allowing to collect repeatable results.

To complement the thermal controller with a power-performance one in order to perform more representative experiments, we implemented a simple power-performance policy setting the DVFS to a filtered value of the load. The actuation values of the thermal and power-performance policies are combined as described in Sect. 5.5.3.

The event-based controller has been implemented as a userspace C++ program, reading the temperature and performing the DVFS actuation via the MSR [6] interface offered by Intel processors. The choice of using the MSR interface allows fast and low-jitter sensing and actuation. The choice of designing the controller in userspace conversely allows to conveniently log the operation of the controller, including the events, temperatures and DVFS actuations, although at the price of a higher overhead. As it is not possible to implement the event generation state machine in hardware on a commercial processor, it has been emulated in software. Every 5 ms, the temperature sensors are read, and a software state machine decides whether or not to call the controller. Finally, though the processor has a per-core temperature sensor, there is only one DVFS actuator for all cores. For this reason, a single loop is used, using the instantaneous maximum among all four temperatures as the controlled one.

5.6.1   Control Quality and Overhead

The first test is aimed at assessing the ability of the proposed controller to keep the temperature at or below the prescribed set point. For this test, the cpuburn application [8] is launched on all cores. This application is specifically designed to maximise the processor current consumption. To test the response to fast power transients and provide a more diverse load profile than a constant power, cpuburn is periodically stopped and restarted. Of course, such a load is not representative of the normal operation of a processor, but step-like *stimuli* are well suited to compare controllers by transient examination.

In this test, the temperature set point has been set to 90 °C, and Fig. 5.10 reports the results. The top plot shows the controlled temperature, the centre plot the DVFS actuation and the bottom one the CPU load. When the measured temperature is lower than the limit, the power-performance policy prevails, and frequency

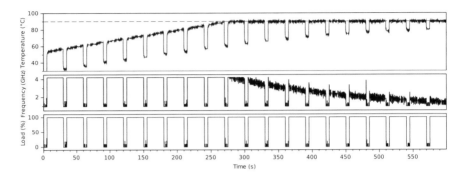

Fig. 5.10 Experimental test 1—controlled temperature and threshold (top), DVFS command (centre) and computational load (bottom)

follows the load pattern, saving power in idle periods. When the cpuburn applications are started and stopped, fast temperature transients, up to $30\,°C/100\,ms$, are observed, while the slow rising trend is due to the heat sink warming up. Once the temperature reaches the prescribed set point, the thermal controller starts to reduce the frequency, and its action becomes stronger over time to compensate for the heat sink temperature trend. The controller successfully keeps the temperature at or below the prescribed limit, with an average of only 14 events per second.

5.6.2 Comparison with the State of the Art

The proposed controller has been compared against thermald [5], a service for the Linux operating system to control the CPU temperature, recently developed by Intel; thermald is configured through a file where it is possible to set the limit temperature. Internally, it uses a PID controller calibrated at start-up not to require control knowledge on the part of the user.

The first benchmark used for comparison is based on cpuburn as the previous test. In this case, cpuburn is run continuously for 260 s, to obtain a constant CPU power. After that, it is periodically stopped and restarted to introduce power transients.

Figure 5.11 shows the results. Both controllers perform acceptably, although thermald causes temperature fluctuations with an amplitude of $10\,°C$, even when the CPU power is almost constant. The proposed controller limits such fluctuations to $+2/-3\,°C$ and is limited mainly by the sensor noise. The proposed controller can achieve this level of performance in controlling the temperature at just 22 events/s. When power transients are introduced, both controllers can prevent temperature peaks.

The second benchmark used is taken from the Intel optimised LINPACK parallel suite [4]. The chosen benchmark consists in solving a system of linear equations,

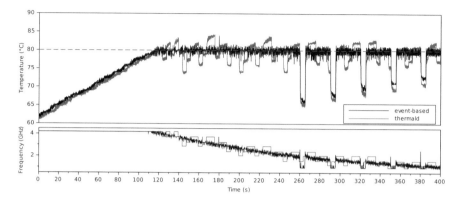

Fig. 5.11 Experimental test 2—comparison between the proposed controller and `thermald` when running `cpuburn`. Controlled temperature and threshold (top) and DVFS command (bottom)

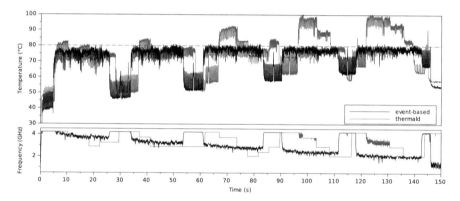

Fig. 5.12 Experimental test 3—comparison between the proposed controller and `thermald` when running `LINPACK`. Controlled temperature and threshold (top) and DVFS command (bottom)

and stresses the CPU floating point pipeline considerably. This test is relevant because `LINPACK` alternates execution phases that cause very different cache miss rates; thus, the CPU is never idle, which means 100% *load* throughout the test, but *power* varies significantly.

As a consequence, see Fig. 5.12, the two controllers yield significantly different results. The event-based one successfully keeps the temperature set point. The only adverse effect of the mentioned power bursts is an increase in the number of thermal events, on average 61 per second. `thermald` is instead simply not capable of keeping the temperature below the limit, due to its slow actuation. It succeeds in the first two iterations, when the heat sink is warming up, but when the slow temperature trend has raised near enough to the limit, it fails to react timely, and the CPU temperature reaches 100 °C between 96 and 98 s from the experiment start, and again between 121 and 131 s. In those time frames, the hardware thermal protection is engaged, as shown by the evident high-frequency content of the DVFS command.

As a final comment, while `cpuburn` executes its task in an endless loop until terminated, the `LINPACK` benchmark performs a finite amount of work, and thus the completion time can be used as a metric of the CPU performance. During the first and second iteration, when both controllers succeed in keeping the temperature limit, the event-based controller results in an increased computational speed, finishing the second iteration after 53.5 s compared to 58 s with `thermald`—a 7.8% improvement. During the following iterations, the execution is slightly faster with `thermald`, but this is because it exceeds the selected temperature limit, and not negligibly.

Summarising, the proposed event-based scheme is capable of controlling temperature at the speed required to counteract fast transients, with all the advantages of software configurability and openness to new control laws, with full power/performance integration, and with a very low overhead.

5.7 Conclusions

In this chapter, we presented the event-based digital controller for the thermal management of microprocessors that has been developed as part of the HARPA project. Peculiar of our proposal is the integration into a unitary scheme of temperature control and power/performance trade-off. We carried out a stability analysis, and set preliminaries for a performance one. The proposed controller was tested and assessed by means of a comprehensive Modelica library, that we created by extending previous works, and can serve as a solid benchmark, open to multi-physics settings, for control studies on the matter addressed herein. We also showed some experimental results obtained on a modern processor architecture and compared our solution to the state of the art, in the form of a recently proposed operating system thermal daemon to be used jointly with classical power/performance governors.

Our proposal behaves significantly better than the available alternatives and is easy to set up, calibrate and maintain. We can thus state that the presented control system acts as a technology enabler by allowing the safe operation of high-density processor, also for what concerns their fastest thermal dynamics. As a final note, concerning the proposed control solution, a PCT (Patent Cooperation Treaty) application has been filed.

References

1. Casella, F., & Leva, A. (2006). Modelling of thermo-hydraulic power generation processes using Modelica. *Mathematical and Computer Modelling of Dynamical Systems, 12*(1), 19–33.
2. Dennard, R., Gaensslen, F., Rideout, V., Bassous, E., & LeBlanc, A. (1974). Design of ion-implanted MOSFET's with very small physical dimensions. *IEEE Journal of Solid-State Circuits, 9*(5), 256–268.
3. Esmaeilzadeh, H., Blem, E., Amant, R. S., Sankaralingam, K., & Burger, D. (2012). Dark silicon and the end of multicore scaling. *IEEE Micro, 32*, 122–134.
4. Intel Corporation. Intel optimized LINPACK benchmark.   https://software.intel.com/en-us/articles/intel-math-kernel-library-linpack-download
5. Intel Corporation. Linux thermal daemon. https://01.org/linux-thermal-daemon
6. Intel 64 and IA-32 Architectures Software Developers Manual. http://www.intel.com/content/www/us/en/architecture-and-technology/64-ia-32-architectures-software-developer-system-programming-manual-325384.html
7. Leva, A., Terraneo, F., & Fornaciari, W. (2015). Event-based thermal control for high-density processors. In *Proceedings of 1st International Conference on Event-based Control, Communication, and Signal Processing*, Krakow (pp. 1–8).
8. Redelmeier, R. The `cpuburn` CPU stress tester.   http://manpages.ubuntu.com/manpages/precise/man1/cpuburn.1.html
9. Taylor, M. (2013). A landscape of the new dark silicon design regime. *IEEE Micro, 33*(5), 8–19.

Part III
HARPA-RT Layer

Chapter 6
HARPA RT

**Dimitrios Rodopoulos, Nikolaos Zompakis, Michail Noltsis,
Francky Catthoor, and Dimitrios Soudris**

6.1 HARPA RT

6.1.1 Primitives

As technology nodes approach deca-nanometer dimensions, a variety of reliability violations threaten the binary correctness of processor execution. Computer architects typically enhance their designs with mechanisms that correct such errors, possibly at the cost of extra clock cycles. However, this clock cycle overhead increases the performance vulnerability factor (PVF) of the processor. The goal is to minimize the aforementioned cost using dynamic voltage and frequency scaling (DVFS), which is typically featured in the state-of-the-art processors. Towards this direction, the concepts of deadline vulnerability factor (DVF), cycle noise, and slack are introduced. A closed-loop implementation is proposed, which configures the clock frequency based on the observed slack value. Thus, the temporal overhead of error mitigation is absorbed and performance dependability is improved. We evaluate the transient and steady-state behavior of our approach, proving its responsiveness within less than 1 ms for a wide variety of error rates.

Before presenting the details of the HARPA RT, we will establish a common theoretical background, introducing all key terms. Then, the two components of the HARPA RT with System Scenarios are presented: (1) a PID Controller, which ensures stable control of the performance dependability scheme and (2) a System

---

D. Rodopoulos (✉) · F. Catthoor
Imec, Leuven, Belgium
e-mail: drodo@microlab.ntua.gr; catthoor@imec.be

N. Zompakis · M. Noltsis · D. Soudris
MicroLab-ECE-NTUA, Athens, Greece
e-mail: nzompakis@microlab.ntua.gr; mnoltsis@microlab.ntua.gr; dsoudris@microlab.ntua.gr

© Springer International Publishing AG, part of Springer Nature 2019
W. Fornaciari, D. Soudris (eds.), *Harnessing Performance Variability
in Embedded and High-performance Many/Multi-core Platforms*,
https://doi.org/10.1007/978-3-319-91962-1_6

Scenario component, which detects the applicable performance dependability target at all times. The latter is very important, since it indirectly provides information to the HARPA RT about the extent of performance variability that needs to be mitigated.

Error mitigation inflates the number of clock cycles that is required to complete portions of the instruction stream (corresponding to TNs), which pass through the processor. In order to guarantee dependable performance, this overhead should be quantified and absorbed. We formulate this problem using the following definitions. From prior art, we repeat that the concept of the PVF is the per unit increase in the number of clock cycles required for a specific instruction stream, as a result of temporal error mitigation overheads [2].

Definition 1 **Cycle budget** (N) is the number of clock cycles (cc) of relevant computation assigned to each TN. This can be determined at design time or at runtime according to the following equation, based on a specification for the cycles per instruction (CPI) of the processor, the number of instructions (L) in a TN, and a user-defined PVF tolerance ($\text{PVF}_{\text{limit}}$) :

$$N = (1 + \text{PVF}_{\text{limit}}) \times \text{CPI} \times L \tag{6.1}$$

Definition 2 **Cycle noise** (x) is the sum of clock cycles (cc) that are inevitably wasted by mechanisms like ECC, shadow latches, etc. In general, it comes across as a result of error mitigation, which consequently leads to PVF degradation.

Definition 3 **Frequency multiplier** (x) is a positive real number by which the default clock frequency (f) is multiplied, reflecting possible DVFS configurations.

Definition 4 **Deadline vulnerability factor** (**DVF**) is the per unit difference between the real execution time (T_{real} with an m_{real} frequency multiplier), including cycle noise, and a reference execution time (T_{ref} with a frequency multiplier m_{ref}), where no cycle noise occurs:

$$\text{DVF} = 1 - \frac{T_{\text{ref}}}{T_{\text{real}}} = 1 - \frac{\frac{N}{m_{\text{ref}} \times f}}{\frac{N+x}{m_{\text{real}} \times f}} = 1 - \frac{m_{\text{real}} N}{m_{\text{ref}}(N + x)} \tag{6.2}$$

Let us assume that a stream of TNs is executed by a processor. For each TN, a frequency multiplier can be selected ($m[n]$). A cycle noise value ($x[n]$) is also associated with each TN. A recursive formulation of DVF at the end of each TN is possible. An expression is derivable from the DVF definition, by expressing $T_{\text{real}}[n]$ using $T_{\text{real}}[n-1]$. In that sense, we can write:

$$\text{DVF} = 1 - \frac{T_{\text{ref}}[n]}{T_{\text{real}}[n]} = 1 - \frac{T_{\text{ref}}[n]}{T_{\text{real}}[n-1] + \frac{N[n]+x[n]}{m[n]f}} = 1 - \frac{T_{\text{ref}}[n]}{\frac{T_{\text{ref}}[n-1]}{1-\text{DVF}[n-1]} + \frac{N[n]+x[n]}{m[n]f}}$$

$$\tag{6.3}$$

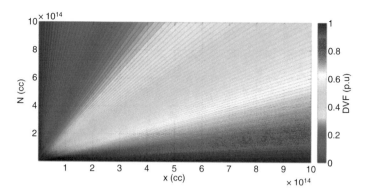

Fig. 6.1 Deadline vulnerability factor (DVF) converges to zero, by definition, when $x \ll N$

Understandably, this formulation is complicated and has increased computational requirements to derive (many math operations). Even if compatible to the state-of-the-art formulation (e.g., PVF), its complexity is prohibitive for a run-time engine that is supposed to configure a platform and guarantee performance dependability within roughly 1 ms. As a result, we need a recursive metric that requires few math operations, thus enabling hardware integration of a DVF degradation monitor.

It is also very interesting to observe the long-term trends of DVF according to the above definition, assuming that no DVFS-like countermeasure has yet been engaged. In Fig. 6.1, we present various values of DVF for a sweep of N and x and identify certain regions of interest:

- When $x \gg N$, DVF converges to 1, reflecting processing entirely dominated by cycle noise.
- Another region is where $x \approx N$, where we observe intermediate values for DVF.
- The upper-left side of Fig. 6.1 corresponds to $x \ll N$ and reflects realistic cases, since cycle noise is not expected to exceed cycle budget. We see that as N increases (e.g., when useful processing follows a cycle noise burst), DVF converges to zero. It is to be expected that, as long as no cycle noise occurs, DVF will asymptotically converge to 0. Thus, deadline vulnerability is restricted to a significant portion of the execution, right after cycle noise has manifested itself.

In Fig. 6.2, we see a pseudo-transient evolution of DVF, assuming impulse cycle noise occurrences. Again, we observe that DVF asymptotically converges to zero, immediately after an impulse of cycle noise has been injected. This makes DVF very suboptimal for on-the-fly performance variability depiction, since it always requires calculations from the initial point of execution. The situation is made even worse, given the math operation requirements of DVF. Hence, a new term is needed to provide at all times the proximity of the system to a deadline violation.

Definition 5 Rate Divergence Slack (s) expresses the divergence of the processor from a default clock budget that is assigned to a TN ($N[n]$) assuming the default

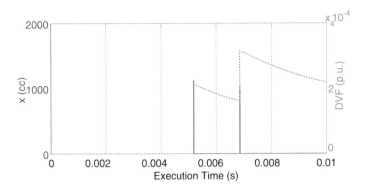

Fig. 6.2 Transient view of DVF, for impulse injections of cycle noise

Table 6.1 Operations required for performance variability indicators

	Divisions	Multiplications	Additions	Subtractions
Deadline vulnerability factor	3	1	2	2
Rate divergence slack	0	0	1	2

(m_{ref}) and current (m_{real}) frequency multipliers. It is given, by the following Equation, whereas for the sake of brevity, we will refer to this term simply as slack for the rest of the current document:

$$s[n] = \frac{N[n]}{m_{\text{ref}}} - \frac{N[n] + x[n]}{m_{\text{ref}}} + s[n-1] \tag{6.4}$$

It appears that, by definition (see Table 6.1), slack is a simpler metric when compared to DVF, in terms of required math operations. Considering the slack evaluation, we have reduced the cost significantly because the variable multiplications and divisions are simplified and only a few shifts and add/subs are left. If we are looking at a HW or firmware implementation of the HARPA RT, it is very important to consider such overheads.

It is also important to show that *iff slack is equal to zero, then DVF is also equal to zero* [9]. For the sake of brevity, we do not elaborate on the proof of the above material equivalence. However, given the above recursive definitions, the proof can be easily derived using mathematical induction and assuming that $s[0] = DVF[0] = 0$, which can be expected at the beginning of a TN series execution. Having provided an illustration of the proof, we can conclude that aiming at $s[n] = 0$ is equivalent with aiming at $DVF[n] = 0$, which is that ultimate goal of mitigating performance variability. That way, it is consistent to design the HARPA RT in such a way that processor slack is forced to converge to zero.

6.1.2 Performance Variability Model

In this subsection, we will illustrate how performance variability is perceived in the context of the HARPA RT. Cycle noise is the key term for the defined model, given that it quantifies the intervention of error correcting mechanisms with the processor performance. We define the rate at which cycle noise occurs and its peak amplitude. If a detected and corrected error [4] occurs, the correction mechanisms add cycle noise (x) to the TN cycle budget (N). Hence, the rate of cycle noise is described by an MTTF estimation [4]. The probability of cycle noise occurrence after wall-clock time is given by the following equation, which is equivalent to the (MTTF,1) distribution [8]:

$$P = 1 - \exp\left(\frac{-\Delta T}{\text{MTTF}}\right) \tag{6.5}$$

In a simulation environment, a random number r can be created and compared to P. If we assume that cycle noise is injected at that point of the execution and that it is equal to, as shown in Fig. 6.3. The parameters μ and σ depend on the PVF impact of error correction and reflect the amplitude of cycle noise. As a result, in a simulation environment, we can assume the injection of cycle noise bursts, which generally follow the above model. Through the rate and amplitude parameters, one can adjust the aggressiveness of performance variability. Also, such a model is linkable to reliability models, if one assumes the error rates of architectural components, as well as the error correction coverage that is applied to the processor.

The aforementioned model definition further defines the interface of the HARPA RT with the greater HARPA workpackage on performance variability modeling. In that sense, the outcome should be compatible to the above formulation. That way, in order to verify the HARPA RT in a prototype or a simulation environment (regardless of the underlying scenario methodology), the injected cycle noise should

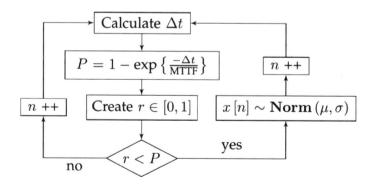

Fig. 6.3 Modeling of cycle noise during TN execution

come about as a result of processor variability that is corrected by specific circuitry, which in turn adds temporal overheads (causing the PVF to increase).

6.1.3 HARPA RT System Scenarios

The HARPA RT utilizes a PID controller to mitigate the performance degradation effects due to RAS interventions. The PID controller operates on cycle budgets that a scenario detector provides. Thus, the definition of the system scenarios is critical for the optimal responses of the HARPA RT. Figure 6.4 depicts the functionality of the proposed PID controller. Table 6.2 explains the variables of the proposed instantiation.

The proposed control scheme (for the case of constant cycle budget) is shown in Fig. 6.4. The idea is to create an error signal ($e[n]$) that indicates the currently experienced slack and adjust the frequency of the processor accordingly. This error signal is amplified in traditional discrete-time PID fashion, by adjusting the respective gains, according to the following equation:

$$\hat{m}[n] = \underbrace{e[n]k_p}_{\text{proportional component}} + \underbrace{\hat{m}_i[n-1] + e[n]k_i}_{\text{integral component}} + \underbrace{(e[n] - e[n-1])k_d}_{\text{differential component}} \quad (6.6)$$

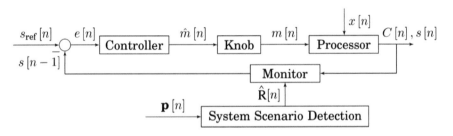

Fig. 6.4 Block diagram of the proposed performance dependability scheme

Table 6.2 Variables of the proposed HARPA-RT instantiation

$s_{ref}[n]$	Target slack specification (0 unless else specified)	$x[n]$	Timing noise interfering with RTS
$s[n-1]$	Slack at the end of the previous RTS	$C[n]$	Actual budget of current RTS
$e[n]$	Error signal $s_{ref}[n-1] - s[n-1]$	$s[n]$	Slack at the end of the current RTS
$\hat{m}[n]$	Proposed frequency multiplier	$\hat{R}[n]$	Default timing budget for current RTS
$m[n]$	Applied frequency multiplier	$p[n]$	System RTS parameters

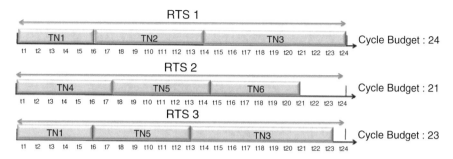

Fig. 6.5 Thread nodes sequences into run-time situations (RTSs)

The PID controller reacts at each RTS consisting of thread nodes (TNs). The entire program code is divided up into TNs. Each TN is a sequence of instructions with deterministic behavior (e.g., the body of a loop) that is executed in an atomic way. Run-time situation (RTS) is a term that is assigned to a TN or a sequence of TNs (see Fig. 6.5), in terms of the TNs timing (latency) or energy budget. Resolution of this label assignment is based on parameters of the TN, which are called RTS parameters. Due to the existence of multiple TNs in applications that are executed with several combinations (see Fig. 6.5), the RTS with multiple TNs allows the PID controller to respond to a more coarse granularity.

For dependable performance to be realized, each RTS needs to respect a timing deadline as indicted by its default cycle budget and the applicable reference frequency (m_{ref}). Due to the variety of RTSs that appear in real applications, it is rather suboptimal to store their respective cycle budgets, let alone detect them at runtime. In other words, the number of RTSs that are encountered in the instruction stream of a processor is quite complex to handle. Apart from storing the cycle budgets per RTS, it is additionally complex to detect the RTSs. The system scenario approach [1] overcomes this issue by clustering RTSs into cumulatively representative cases called system scenarios (or simply, scenarios). Assigning a cycle budget that is disproportionately high to an RTS may lead to degradation of the system performance, due to the relaxation of the timing deadline that is assigned. Conversely, an underestimated cycle budget for an RTS may create a false impression of negative slack, thus pushing the PID controller to an over-boosting that will increase energy consumption. System scenarios aim [10] to balance the above trade-off by clustering RTSs to scenarios ensuring the worst-case RTS for each scenario without losing significantly in terms of accuracy (achieved precision around 90%) Assuming a workload that is composed of A different RTSs, each with cycle budget Ni, where $i = 0, 1, A$, the goal of this methodology is to narrow down to B scenarios, each with cycle budget estimators (or simply, scenarios) $\hat{N}_j$ where $j = 0, 1, B$ and $A >> B$. The reduced number of scenarios to be considered (in comparison to RTSs) facilitates their storage in the system and their less complex utilization at runtime. Additionally, clustering from cycle budgets N_i to estimators $\hat{N}_j$ should be such that no cycle budget is excessively under- or overestimated.

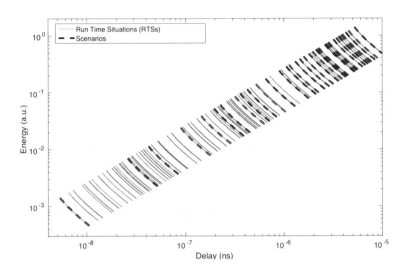

Fig. 6.6 RTS clustering to scenarios on a representative spectrum sensing application —Energy is calculated using the square law for dynamic power, i.e., $P_{\mathrm{dyn}} \infty f V_{dd}^2$ and numbers are normalized to the smallest RTS at $V_{dd} = 1.2$ V and $f = 2.588$ GHz

To achieve this, the difference between N_i and $\hat{N}_j$ has to be minimized for every clustering. We illustrate the above concepts with an example: Fig. 6.6 illustrates an example of a significant number of RTSs of a real spectrum sensing application that have been clustered to scenarios. Each RTS represents a certain cycle budget. Given a number of OPP points on any processor that supports DVFS, we can draw an energy vs. delay line, similar to the solid lines of Fig. 6.6. For these very lines to be produced, we are using the OPP points of a well-documented processor [6] which have also been used in the previous work [7]. The multitude of solid lines in Fig. 6.6 corresponds to a spectrum sensing application and already represents the complexity of handling TNs using the RTS abstraction. Clearly, this amount of information needs to be abstracted further. System scenarios can replace the many RTSs with a more manageable set of scenarios, to be used as cycle budget estimators. The idea is to bin RTSs, based on similarity of their cycle budgets as they are projected on the multidimensional Pareto metric space (e.g., Energy Delay). In that sense, RTSs that require a similar number of cycles to execute are binned to the same scenario. The scenario is uniquely identified with its cycle budget estimator $\hat{N}_k$, which is the minimum cycle budget of the RTSs that have been binned therein. By extension, TNs that belong to similar RTSs are also falling under the same scenario when perceived at runtime. More specifically, in Fig. 6.6, we see an artist's illustration of the clustering process. The assigned cycle budget is that of the fastest RTS in the scenario. As we move away from the origin of the axis, new scenarios are defined and certain RTSs are mapped to them. The assigned cycle budget is that of the fastest RTS in the scenario. Eventually, the wealth of RTSs is produced by all possible combinations of application configuration such as accuracy settings and input size. The system scenario methodology reduces this huge set of RTSs to a more manageable set of scenarios. Given that the fastest RTS is dictating the budget estimator of each scenario, it is clear that the complexity reduction detailed above is introducing an accuracy overhead, which is typically called clustering overhead. There lies a trade-off between the accuracy and the number of the derived scenarios.

A fine grain clustering decreases the gap between the budget estimators of the same scenario but increases the number of the scenarios and vice versa. Assigning the minimum cycle budget from the RTSs that belong to a scenario cluster forces the rest included RTSs due to their distance from the worst case to be more restricted, compared to their actual timing deadlines. This trend is proportional to the clustering overhead that leads the PID controller to act more aggressively, boosting the operation frequency. This effect is not necessarily undesirable for the purpose of our design. The mission of the PID controller is to absorb the interfering cycle noise of the RAS interventions. The extra frequency boosting operates preventively to the extra delays imposed by cycle noise. The clustering overhead acts as an offset that permits to partially eliminate the introducing noise without increasing slack. This permits less PID controller interventions preventing the repeated frequency DVFS switches. As a result, the inevitable clustering overhead is not undesirable, provided that it does not force the PID controller to "worst-case behavior" (i.e., forcing it to suggest the maximum operating frequency). Without the loss of generality in our claims, a simple linear binning scheme, which results in the cycle budget estimator, is shown in Fig. 6.6. Prior art already features many ways to cluster scenarios at design time and detect them at runtime from the RTSs encountered during TN execution (Fig. 6.14).

6.1.4 Different Classes of Scenarios

The PID controller as refereed previously absorbs the interfering cycles of the RAS interventions. The controller adjusts the processor frequency, actuating suitable DVFS schemes. The incoming RTSs are valued at a certain cycle budget that is used as reference for the expected operation cycles. Cycle noise being superimposed on cycle budget introduces negative slack that pushes the PID controller to a new DVFS boosting. The open issue is how the PID controller can act proactively. Due to the RAS intervention randomness, the deployment of prediction mechanisms that fully forecast the interfering noise at runtime is neither realistic nor cost effective. The key point is how to encapsulate in the reference cycle budget the gradual and partly predictable performance of the platform. The goal is to provide a low-cost run-time cycle budget mechanism that will provide opportunities for less slack without eliminating it.

The existing slack reclaiming mechanism exploits a system scenario detector. The system scenarios provide the default cycle budget of the running RTSs. System Scenarios are clusters of RTSs with similar cycle budgets as shown previously in Fig. 6.6. The hierarchy and the structure of the RTSs in the System Scenarios is the key for an efficient PID implementation. The effects of hardware degradation increase RAS interventions so a static definition of system scenarios and an equally static cycle noise formulation is not expected to yield dependable performance in an efficient way. The dynamic adjustment of the system scenario concept to the new requirements is an intermediate step towards the ultimate introduction of dynamic scenarios.

Outlining the system scenario primitives, RTSs are discrete executions of TN sequences, with a default cycle budget while the most restricted deadline into a scenario represents the fixed cycle budget for the entire RTSs. Due to variability of the RTSs into the same scenario, overestimated constraints push the PID controller to a frequency multiplier higher from what is required, thus partially mitigating noise cycles and paying a price at energy consumption. A noise shift higher than a limit creates negative slack and the PID controller is called to force to zero with a frequency increase. On condition that the RTS positions into scenarios can be updated at runtime, the scenario detector can modify the RTS hierarchy by re-clustering the existing scenario distribution. This process provides a run-time adjustment of the RTS cycle budget estimations [1]. This leads the PID controller to act partially proactively (see Fig. 6.7a) considering RAS events that are gradually becoming a norm on the target processor. To ensure a fully proactive operation [5] in a complete dynamic scenario approach, a predictor has to provide a runtime trend of the incoming noise. Redistributing the predicted slack across a future scheduling period in a globally optimized way as pictured in Fig. 6.7b we will achieve an even better trade-off between delay and energy. In the context of the current chapter, we focus on the scenario adaptivity through re-clustering defining the conditions of the

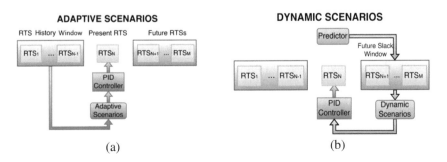

Fig. 6.7 Flow charts adaptive vs dynamic scenarios. (**a**) Adaptive scenarios. (**b**) Dynamic scenarios

scenario modifications at runtime. The incorporation of a predictor is not part of the current work and represents an individual challenging issue that needs more elaborate research that will be part of the future work.

6.1.5 Adaptive Scenarios

In this subsection, we explain how an adaptive re-clustering mechanism is built from scratch. The objective is to keep the operation cost of such a mechanism low, adapting at runtime modifications that keep our system updated. The key issue is the splitting of the processes that take place at design time and at runtime. The concern is to build a light consuming flexible mechanism that reacts quickly ensuring the respecting deadlines. Thus, the heavy computation parts are pushed to the design time, keeping at run-time decisions with short response time. This presupposes a design time analysis that builds scenario metadata without requiring repeating the whole calculations at runtime.

In this direction, the involved TNs of the targeted application are analyzed at design time. The targeted application is split into TNs based on specific criteria [3]. The combination of TN execution creates the exploration space of the involved RTSs. To identify the operation cost of each RTS, we analyze the several trade-offs for the available DVFS knobs. This information is critical to define which knobs deal with the application requirements. The depicture of the RTS on a Pareto space provides a good representation of the RTSs behavior and a comparison of the interested design points presents the cost impact of each RTS. This comparison creates a hierarchy of the RTSs based on their position in the metric space (see Fig. 6.8). Based on this hierarchy, cluster domains define the enclosed RTSs into scenarios. Thus, a splitting of the metric space into domains is a first approach that provides an abstraction view of the operation cost hiding the details of the individual RTSs. The split point is defined as the upper bound (see in Fig. 6.8 RTS1 red

Fig. 6.8 RTS Pareto curve hierarchy into scenarios

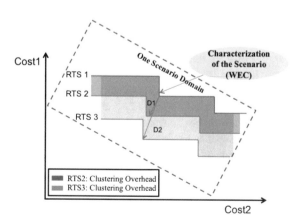

curve) of each scenario that distinguishes it from the other scenarios. Two criteria specify these points. The first criterion is that the chosen upper bound has to have the minimum distance from the included RTS Pareto curves (see in Fig. 6.8 D1, D2). The second criterion that cannot be visualized is the clustering of the suitable combinations of RTSs that minimizes during the operation of switching between the scenarios. Thus, in the first case an algorithm has to solve a trivial distance minimization problem and a second algorithm solves a probability distribution issue. Both algorithms tend to be competitive in respect to the targeted improvements. More precisely, the first algorithm aims to reduce the clustering overhead by increasing the switching overhead and vice versa. The achievement of the minimum total overhead is a balance between them, and more details can be found in [10]. Both algorithms can be highly computation consuming due to the involved number of RTSs and the derived scenarios so they take place at design time and provide the necessary information for the run-time stages. In particular, the final clustering is based on providing information (metadata) about the cycle budget of each scenario based on the most time-consuming RTS that will be exploited at runtime.

At runtime, a re-clustering mechanism identifies the running RTS and the correlated slack. If the same RTSs present cycle noise repeatedly, this means that these RTSs require a regrouping adjustment. For this reason, we use a sliding window that traces the history of the RTS running. The re-clustering starts by priority from RTSs with the most significant slacks at more frequent times. The RTS slacks are watched on fixed time windows. The result of the extra cycle noise is to increase the overhead of the effected RTS pushing it in a more costly position at the metric space (delay-energy). The possible transitions of an RTS due to the added cycle noise are outlined in Fig. 6.9a–e. RTS1 represents the shifted-RTS. For each repositioning, an individual re-clustering decision (each decision will be described at the following subsection) is taken to equalize the performance degradation. The scenario metadata that were initialized at design time are updated after a re-clustering. Apart from the scenario budget, information is included about the distance from the neighboring scenarios. This information permits the rapid recognition of the optimal re-clustering decision updating only the scenario data that are changed. Each re-clustering decision is taken based on the new position of the RTS at the metric space after a reviewing of the updated data, ensuring that the RTS deadlines will be respected with the minimum cost. The re-clustering is a parallel process that does not interrupt the operation of the PID controller. In any case, the re-clustering time can be easily kept a fraction of the RTS operation time, due to the simplicity of the algorithm to find the new shifted RTS position without reordering the whole RTSs by scratch. At the end of each re-clustering, the PID controller has updated information about the suitable DVFS scheme that minimizes the generated slack paying a price in energy. A trade-off between the produced slack and energy exists that will be explored in the last section with experimental results.

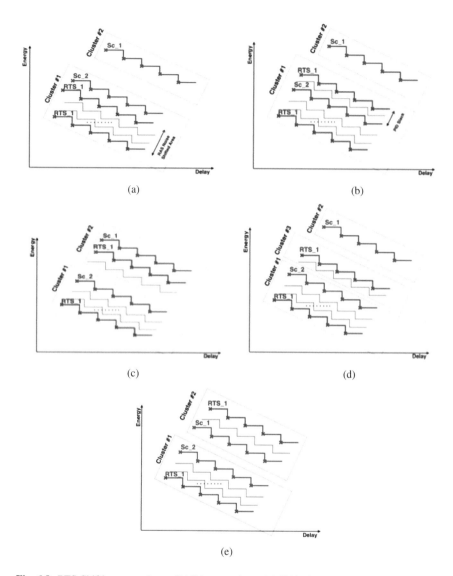

Fig. 6.9 RTS-Shifting cases due to RAS interventions. (**a**) Shifted-RTS1 remains at the scenario bounds. (**b**) Shifted-RTS1 remains at the scenario changing the bounds. (**c**) ShiftedRTS1 is merged by another neighboring scenario. (**d**) ShiftedRTS1 is creates an individual scenario. (**e**) ShiftedRTS1is merged by another scenario changing the new scenario bound

6.1.5.1 Adaptive Scenarios Instantiated for the Cycle Noise Reduction in the PID Controller

Concentrating on the re-clustering decisions, the key point, as referred previously, is to distinguish transient from (quasi-)permanent cycle noise events. For example, a very brief cycle noise burst that is never reoccurring should not be triggering the re-clustering phase. On the other hand, when cycle noise bursts appear to have a more predictable behavior, re-clustering RTSs may be required in order to increase the efficiency of our PID-controlled performance dependability scheme. The hardware degradation effects occur during the operation life and cannot be predicted in advance. Thus, an updated trace history of the budget and noise per RTS is needed to consider their fluctuation with time. Identifying a new trend in cycle noise manifestation is important for the changes/reformulation of the system scenarios through RTS re-clustering.

The RAS interventions change the RTS hierarchy into scenarios by removing from the default cycle budgets the noise cycles and shifting the position of the RTSs in the scenarios. Thus, more restricted deadlines trigger more aggressive DVFS schemes that absolve the extra cycles. As the noise is impossible to be accurately defined at runtime, the negative slack is identified as noise. Examining the re-clustering decisions that have to be considered at runtime, we outline the following cases:

1. In the first case, despite the added cycle noise, the scenario budget is restricted enough and no slack exists, hence there is no need for scenario change. This case is visualized by cost perspective in Fig. 6.9a. The slack absence is visible from the fact that the added noise does not exceed the distance between the RTS and the scenario budget, exploiting the clustering overhead as an offset. Thus, clustering overhead can be utilized as a proxy of cycle noise tolerance. A high clustering overhead permits wide RTS shifting without slack but with negative effects at the usage of overestimated DVFS schemes. The next four cases cover situations where the negative slack cannot be avoided.

2. In the second case, the noise pushes an RTS to a more restricted budget than the entire scenario budget. The new RTS position forces the revision of the existing scenario budget (see Fig. 6.9b). The new budget is decreased per equal cycles to negative slack.

3. The differentiation at the third case is that another existing scenario merges the shifted RTS (see Fig. 6.9c). The new position of the RTS fits better to another existing scenario so an updated list changes the distribution of RTS into scenario.

4. The fourth case is when the shifted RTS represents a new distinct scenario (see Fig. 6.9d). Two reasons lead to such a decision: (1) none of the existing scenario cycle budgets cover the new RTS constraint so a new worst-case scenario is needed or (2) the creation of a different scenario is more efficient from clustering overhead perspective, than to be merged by an existing scenario due to the distance between the scenarios.

5. The fifth case is a combination of the second and the third case. The RTS is merged to another existing scenario changing its upper bound (see Fig. 6.9e).

6.1.6 Experimental Results

Having applied static system scenarios and adaptive system scenarios to the RTSs of a representative application [9] (starting up with roughly 30 scenarios), we submit this workload to a simple RTS-level simulator [7], to compare the efficiency of the new concept with the existing system scenario approach. The targeted application is a real spectrum sensing application that consists of 25 TNs that constitute an exploration space of about 800 RTSs (see Fig. 6.10). For the sake of completeness, we repeat the basic functionality of the simulator here. As illustrated in Fig. 6.4, the simulator operates at RTS granularity. For each RTS, the scenario clustering provides the cycle budget estimator $\hat{R}_n$.

Additionally, cycle noise is injected based on the values of Λ and that represent the rate and amplitude of RAS events, respectively. We assume that cycle noise follows a bivariate Gaussian distribution with mean Λ and M values (λ and μ) and sigmas ($\sigma\{\Lambda\},\sigma\{M\}$). The timing and power models of the simulator create traces from which DVF and total energy can be deduced. Given the definition of cycle budget estimators and their derivation, DVF expresses the per unit increase in execution time required for an instruction stream, following the temporal overhead of invoked RAS mechanisms and is formulated by Eq. (6.7). Respectively, PVF expresses the per unit increase in number of clock cycles required for a specific workload, as a result of RAS technique invocation and is formulated by Eq. (6.8). It is noteworthy that all overheads of DVFS actuation are included in both the timing and power models [7]. Finally, at the end of each RTS, the PID controller selects the frequency multiplier of the next iteration, as detailed in Fig. 6.4.

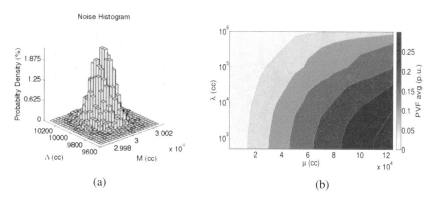

Fig. 6.10 Statistical representations of the simulation conditions. (**a**) Histogram of the cycle noise injected in one RTS steam. (**b**) PVF average values of 100 identical iterations

$$\text{DVF}[n] = 1 - \frac{\sum_{i=0}^{n} \frac{N_i}{m_{\text{ref}}}}{\sum_{i=0}^{n} \frac{N_i + x_i}{m_i}} \tag{6.7}$$

$$\text{PVF}[n] = 1 - \frac{\sum_{i=0}^{n} N_i}{\sum_{i=0}^{n} N_i + x_i} \tag{6.8}$$

The simulation results presented herein correspond to a set of random streams of RTSs, drawn from the scenarios of Fig. 6.6. The simulated workload assumes equally probable RTSs. Cycle noise is injected by sweeping μ and λ and assuming $\sigma\{M\} = 50$ and $\sigma\{\Lambda\} = 10$. The noise histogram for one stream of RTSs is shown in Fig. 6.10a. In fact, each measurement corresponds to a stream of 105 RTSs and, when enabled, PID controller gains are $k_p = 1.5 \times 10^{-7}$, $k_i = 5 \times 10^{-8}$, and $k_d = 4 \times 10^{-8}$. For each combination of μ and λ, we present the average over 100 random measurements to provide insight into the statistical variability of our simulation results.

A first conclusion of our experiments is that apart from the extreme cases of noise (high μ and low λ), the PID controller manages to enable dependable performance, especially considering the increment in PVF (see Fig. 6.10b) that cycle noise injection is causing. In addition, for the average values of Fig. 6.11a, b, we observe that $\text{DVFDyn}_{\text{scen}} \leq \text{DVFSys}_{\text{scen}}$. Except from an area ($\mu > 9 \times 10^{14}$)

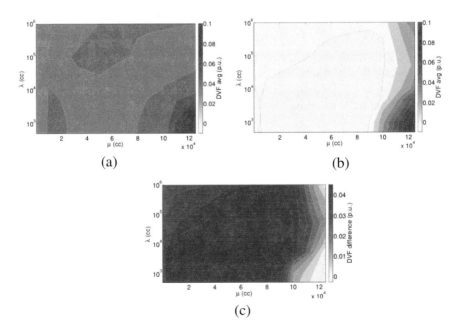

Fig. 6.11 DVF comparison system vs dynamic (adaptive) scenarios. (**a**) DVF system scenarios. (**b**) Dynamic (adaptive) scenarios. (**c**) DVF system scenarios less dynamic (adaptive) scenarios

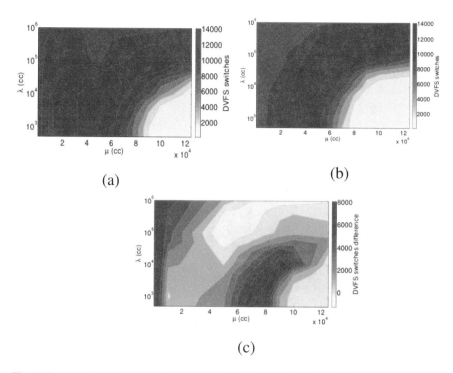

Fig. 6.12 DVFS switches comparison system vs dynamic (adaptive) scenario. (**a**) DVFS switches system scenarios. (**b**) DVFS switches dynamic (adaptive) scenarios. (**c**) DVFS switches comparison system less adaptive scenario

that the mitigation of the cycle noise is not possible due to the limitation at the frequency boosting, adaptive scenarios are evident more effective in terms of eliminating negative slack meeting the deadlines ($DVFDyn_{scen}$) compared to the system scenarios. Another important remark is the deviation in the total number of DVFS switches as shown in Fig. 6.11. It is clear that the adaptive scenarios for the same amount of cycle noise cause significantly less DVFS switches. The scenario adaptiveness seems to counteract the slack fluctuation so the PID controller does not need to overreact (Fig. 6.12).

Finally, total energy (see Fig. 6.13) is increasing as the injected cycle noise profile becomes more aggressive (high μ and low λ). This is to be expected, given that the PID controller needs to work overtime in these cases. The remarkable conclusion is that the deviation between the system and adaptive scenario energy consumption is minor (see Fig. 6.13a, b) and do not exceed the 3.5% at the worst cases (see Fig. 6.13c). This occurs due to the fact that the more energy consuming frequency boosting of the adaptive scenarios is partially equalized by the less DVFS switches (see Fig. 6.12c). Nevertheless, a fundamental trade-off exists between the achieved performance as expressed by the DVF and the energy impact. Figure 6.14 presents this trade-off between static and adaptive scenarios. However, this will also be the

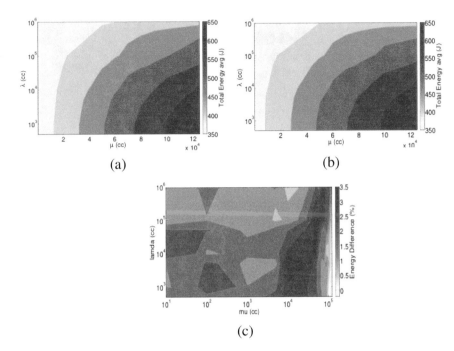

(a)   (b)

(c)

Fig. 6.13 Energy comparison system vs dynamic (adaptive) scenarios. (**a**) Energy system scenarios. (**b**) Energy adaptive scenarios. (**c**) Percentage energy deviation system less adaptive scenarios

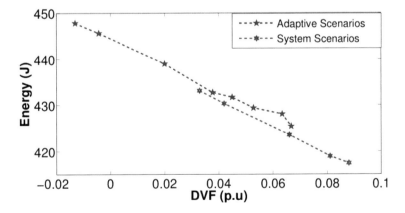

Fig. 6.14 Static—adaptive scenarios performance (DVF) vs energy trade-offs

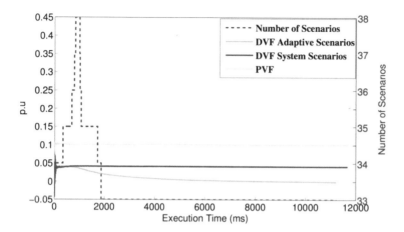

Fig. 6.15 Simulation instantiation ($M: \mu = 30{,}000\sigma = 50$, $L: \mu = 10{,}000\sigma = 100$) of dynamic (adaptive) scenarios applicability on the proposed DVFS scheme for dependable performance

case for all applications where enough DVFS switching options are available to trade-off spending more energy with more slack creation.

The Pareto curves of the two approaches (adaptive vs system scenarios) seem to be very close, presenting similar cost trade-offs (delay-energy). The adaptive scenario appears just a little worse trend (longer distance from the start of the metric axes) due to the extra implementation cost that they require. The gain is their ability to achieve better performance results (DVF close and under zero).

More precisely, Fig. 6.15 highlights how the adaptive scenarios clearly improve the performance dependability at a representative simulation instantiation ($M: \mu = 30{,}000\,\sigma = 50$, $L: \mu = 10{,}000\,\sigma = 10$). While system scenario DVF seems to balance to a static value (0.05), adaptive scenarios adjust DVF close to zero. For a transition period (0–2000 ms), the number of the adaptive scenarios is fluctuating to become to a stable state. During this time, the re-clustering process generates new scenarios increasing gradually their number up to 38 (1000 ms) and after this point some of the existing scenarios are merged to reach the 33 scenarios (2000 ms). Thus, the adaptive process creates and merges scenarios applying the aforementioned re-clustering decisions to find the optimum scenario hierarchy. When this is achieved (after 2000 ms), the scenarios remain stable and the DVF comes close to zero without applying an over-boosting policy (negative DVF) that will consume energy without reason. Thus, Dynamic (adaptive) scenarios seem to improve performance dependability without remarkable energy price.

6.1.7 Conclusion

In the context of the current chapter, we deployed the theoretical background and the operation specifications of the HARPA-RT engine. We presented both a system scenario-based PID mechanism and a partially proactive scenario configuration that adjust the responses of the controller at the presence of RAS interventions ensuring performance dependability. The key point in the adaptive approach is the re-clustering of the existing system scenarios with a flexible way, absorbing the cycle noise that is gradually becoming a norm during the observed application operation. The simulation results prove the efficiency of this approach, ensuring the system dependability constraints while achieving an energy gain up to 15% compared to a guardbanding implementation and exploiting in average 83.9% of the maximally available potential energy gain margins.

References

1. Gheorghita, S. V., Palkovic, M., Hamers, J., Vandecappelle, A., Mamagkakis, S., Basten, T., et al. (2009). System-scenario-based design of dynamic embedded systems. *ACM Transactions on Design Automation of Electronic Systems (TODAES), 14*(1), 3
2. Hardy, D., Sideris, I., Ladas, N., & Sazeides, Y. (2012). The performance vulnerability of architectural and non-architectural arrays to permanent faults. In *2012 45th Annual IEEE/ACM International Symposium on Microarchitecture (MICRO)* (pp. 48–59). New York: IEEE.
3. Ma, Z., Marchal, P., Scarpazza, D. P., Yang, P., Wong, C., Gómez, J. I., et al. (2007). *Systematic methodology for real-time cost-effective mapping of dynamic concurrent task-based systems on heterogenous platforms*. Berlin: Springer Science & Business Media.
4. Mukherjee, S. (2011). *Architecture design for soft errors*. San Francisco: Morgan Kaufmann.
5. Munaga, S., & Catthoor, F. (2013). Systematic design principles for cost-effective hard constraint management in dynamic nonlinear systems. In *Innovations and Approaches for Resilient and Adaptive Systems* (pp. 1–28). Hershey: IGI Global.
6. Park, S., Park, J., Shin, D., Wang, Y., Xie, Q., Pedram, M., et al. (2013). Accurate modeling of the delay and energy overhead of dynamic voltage and frequency scaling in modern microprocessors. *IEEE Transactions on Computer-Aided Design of Integrated Circuits and Systems, 32*(5), 695–708.
7. Rodopoulos, D., Catthoor, F., & Soudris, D. (2015). Tackling performance variability due to RAS mechanisms with PID-controlled DVFS. *IEEE Computer Architecture Letters, 14*(2), 156–159.
8. Xie, M., & Lai, C. D. (1996). Reliability analysis using an additive Weibull model with bathtub-shaped failure rate function. *Reliability Engineering & System Safety, 52*(1), 87–93.
9. Zompakis, N., Noltsis, M., Rodopoulos, D., Catthoor, F., & Soudris, D. (2017). Energy efficient adaptive approach for dependable performance in the presence of timing interference. In *Proceedings of the on Great Lakes Symposium on VLSI 2017* (pp. 209–214). New York: ACM.
10. Zompakis, N., Papanikolaou, A., Raghavan, P., Soudris, D., & Catthoor, F. (2013). Enabling efficient system configurations for dynamic wireless applications using system scenarios. *International Journal of Wireless Information Networks, 20*(2), 140–156.

Chapter 7
Improving Robustness of a Real-Time Spectrum Sensing Application with the HARPA Run-Time Engine

Hans Cappelle, Michail Noltsis, Simone Corbetta, and Francky Catthoor

7.1 Introduction and Motivation

Bit level corruption of digital data is a major reliability concern in microprocessor design. Unfortunately, device downscaling to deca-nanometer dimensions intensifies bit errors through transistor variability [1]. To detect and correct transient or permanent errors in modern microprocessors, designers use reliability, availability, and serviceability (RAS) techniques. A side effect of most RAS mechanisms is a temporary cycle time overhead, resulting in performance variability. In the embedded domain, such variability reduces the dependability for systems with hard quality of service (QoS) constraints. Therefore, in HARPA a cross-layer run-time engine (RTE) mechanism was proposed [2]. The main principle is to absorb performance variability by using knobs such as dynamic voltage and frequency scaling (DVFS). For this purpose, a closed-loop proportional–integral–derivative (PID) controller adapts the processor frequency. Simultaneously, the processor voltage is adjusted to restrict energy use. To validate the HARPA RTE approach, a real-time spectrum sensing application from Thales was selected. This application is elaborated in Sect. 7.2.2 and runs on a quad core ARM® processor platform. In order to induce temperature stress errors, a heat source is used. This chapter is organized as follows:

H. Cappelle (✉) · S. Corbetta · F. Catthoor
Imec, Leuven, Belgium
e-mail: hans.cappelle@imec.be; simone.corbetta@polimi.it; catthoor@imec.be

M. Noltsis
MicroLab-ECE-NTUA, Athens, Greece
e-mail: mnoltsis@microlab.ntua.gr

© Springer International Publishing AG, part of Springer Nature 2019
W. Fornaciari, D. Soudris (eds.), *Harnessing Performance Variability in Embedded and High-performance Many/Multi-core Platforms*,
https://doi.org/10.1007/978-3-319-91962-1_7

- Section 7.2 describes the experimental setup, covering:
  - the measurement setup
  - the sensing application extended with RAS support
  - the implementation choices of the PID controller and the DVFS control strategy
  - the graphical user interface (GUI) to manage and visualize the experiment
- Section 7.3 discusses the measurement results. This consists out of three parts:
  - heat stress-induced errors
  - handling of reliability stress
  - handling of workload stress
- Section 7.4 presents the conclusions of the performed experiments.

7.2 Experimental Setup

The next sections discuss the experimental setup. First, an overview of the selected hardware is shown in Sect. 7.2.1. Generation of temperature stress is supported by means of a hot air source. This can result in functional errors that are not captured by a RAS mechanism on the selected application platform. Therefore, a rollback-based mitigation is emulated in software. This code is combined with the spectrum sensing application, and detailed in Sect. 7.2.2. The implementation of the HARPA RTE PID controller together with selection of DVFS points is detailed in Sect. 7.2.3. Finally, the GUI of the measurement setup is detailed in Sect. 7.2.4. This part also contains the decision logic to detect if a functional error occurred and signals the RAS logic on the platform if such an error was detected.

7.2.1 Measurement Setup

Three main stress factors impact the generation of faults: workload (i.e., instruction and data stream), temperature, and operating voltage [3]. Therefore, the measurement setup should be programmable, and support voltage scaling and temperature variation. Additionally, we need a knob for the parametric mitigation technique of [2]. We selected frequency and voltage scaling to support the HARPA RTE for embedded applications.

Figure 7.1 shows a schematic overview of the selected platform, and Fig. 7.2 shows pictures of the measurement setup parts, which consist out of the following components:

- The Freescale™ IMX6Q board is chosen as a representative embedded application platform. This choice was decided in agreement with Thales. This board

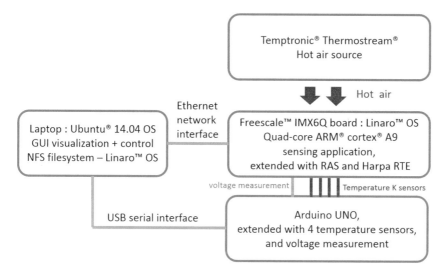

Fig. 7.1 Schematic overview of the measurement setup

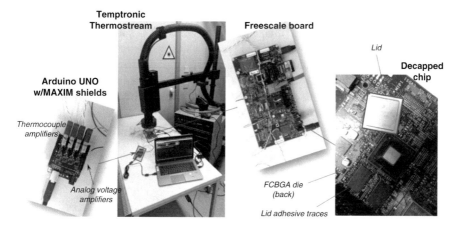

Fig. 7.2 Overview of the measurement setup components

supports DVFS and provides a quad core ARM® cortex® A9 processor core
supporting clock speeds up to 1 GHz. As operating system the Linux® Linaro™
is used, extended with a DVFS software driver.

- A laptop running the Linux® Ubuntu® 14.04, connected to the IMX6Q board
  over Ethernet. The laptop hosts the Linaro™ Network File System (NFS) for
  the IMX6Q board and runs a GUI to control and visualize the experimental data,
  obtained over the Ethernet connection.
- To vary temperature, a Temptronix® ThermoStream® hot air source is used.
  The temperature of this heat source is configurable, and directed at the ARM®

processor. The packaging lid of the processor die is removed to increase the temperature impact on the processor, and to enable direct measurement of the processor temperature.

- To measure temperature, an Arduino UNO extended with a MAXIM shield, supporting four K temperature sensors, is used. The platform reads analog voltages from the thermocouples and from onboard resistor, performs digital to analog conversion, and sends digital traces to the laptop via standard USB interface. Additionally, this board also measures the voltage of the processor core, to estimate its power consumption.

The IMX6Q board offers no visible RAS features to deal with heat stress, and hence these are emulated together with the application.

7.2.2 Spectrum Sensing Application and Functional Reliability, Availability, and Serviceability Emulation

To validate the HARPA RTE for an embedded application, a real-time spectrum sensing application is used. This application has been decided in agreement with our colleagues from Thales TCT, which provided the initial C source code. Spectrum sensing is relevant for software defined radio (SDR), cognitive radios (CR), opportunistic radio access technologies (RAT) dedicated to Frequency White Spaces, and for spectrum monitoring products. Figure 7.3 shows the system context of spectrum sensing. Specifically, the spectrum sensing functionality analyzes a wide spectrum band to detect unused sub-bands. These can be used to allocate frequency bands to a radio protocol, which is achieved by modifying the physical (PHY) and medium

Fig. 7.3 Spectrum sensing system context

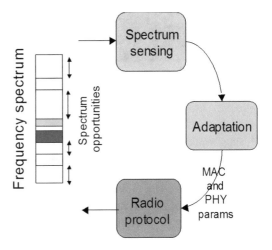

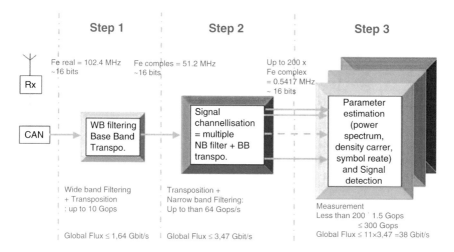

Fig. 7.4 Spectrum sensing complexity

access control (MAC) communication layers. Repetitive sensing of the spectrum is required to gather statistics on the availability of a sub-band.

To evaluate the HARPA RTE, a GSM use case is selected. The workload of this use case exceeds 100 Giga operations per second (Gops/s) and hence is not able to run real time on the selected IMX6Q application board. More powerful platforms were not accessible to us at the time of these experiments, and since optimization of this application was not the objective, the application is run at a lower speed than real time, using stored data as stimuli for the algorithm. Figure 7.4 shows the three main parts of the sensing algorithm:

- Wide band filtering, corresponding to a workload of 10 Gops/s
- Signal channelization with a narrow band filter, corresponding to 64 Gops/s
- Parameter estimation, with a workload up to 300 Gops/s

The reference code of the spectrum sensing application has the following characteristics:

- Three user scenarios with respect to input bandwidth are supported:

  1. Full band, corresponding to an input bandwidth of 102.4 MHz
  2. Half band—51.2 MHz
  3. Quarter band—25.6 MHz

- A sensing resolution bandwidth of 200 kHz is supported in all scenarios, meaning that the input bandwidth is divided into sub-channels with a bandwidth of 200 kHz each for analysis by the sensing algorithm.

Table 7.1 Summary of sensing application parameters

Parameter	Description	Unit	Default value	Min value	Max value
F_s	Sampling frequency	MHz	102.4	25.6	102.4
BW	Input bandwidth	MHz	40	10	40
T_{Buffer}	Buffered signal duration	s	0.1	0.01	1
N_c	Number of filtered samples		200	50	200

- The application is coded in ANSI C, to support ease of porting the solution to different processor architectures.[1]
- To ease hardware requirements, the input and output processing is replaced by file input and output.
- In order to validate functional correctness, a set of reference inputs and outputs is available for each scenario of the GSM spectrum sensing use case.

Table 7.1 summarizes the main parameters of the algorithm. Of particular interest is the buffer size of the algorithm, which corresponds to a duration between 0.01 and 1 s. Since the IMX6Q application board does not contain a RAS mechanism to deal with significant heat stress, it was decided to emulate a functional error mitigation in software. Specifically, a monitor was added which computes as checksum signature to detect function errors, since the input stimuli of the algorithm are known and replayed. If a signature check fails, the application is rolled back to a previous stable point. Data from the application buffer is consumed in fixed steps, and returning to the starting point of the impacted segment allows a safe replay of the segment that was corrupted by a functional error. This procedure leads to extra cycles, or put otherwise a bigger delay. It is the task of the HARPA RTE to compensate this delay by modifying the DVFS setting of the application board. While the implemented RAS mechanism can deal with transient errors, permanent recurring errors result in a system failure. Such failures should be handled by inserting exception handlers at the boot and operating system level and are not covered in the scope of this chapter. To avoid that the reference signature is corrupted, the comparison is done on the laptop, which issues a rollback command if a functional error is detected. This is elaborated in Sect. 7.2.4, which details the GUI.

The sensing application is configured to run on a single ARM® core of the IMX6Q board, so that a single RAS implementation is sufficient for the experiments. The principles of our experiments are however readily transferable to a homogeneous multi-core mapping. Heterogeneous multi-core platforms will require some extensions, especially to handle nonuniform communication between such cores.

---

[1]Which proved to be beneficial when processor architecture change was proposed during the HARPA project.

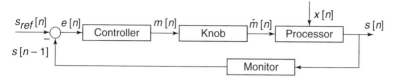

Fig. 7.5 Proportional–integral–derivative (PID) controller block diagram

7.2.3 PID Controller Implementation and DVFS Control Strategy

To control performance, the PID principle is to switch frequency and voltage of the processor based on a reactive response to monitored timing delays (Fig. 7.5). To this end, we have introduced the definition of slack as the timing difference for specific code chunks with deterministic behavior, named as system scenarios [4], between their execution with nominal frequency in the absence of RAS interventions and their execution under fault manifestation. Therefore, negative slack is monitored when errors manifest and RAS events needed to ensure functional correctness are provoked. In this case, the PID controller should decide to boost frequency and voltage in order to ensure performance dependability and convergence of the slack to zero. On the contrary, positive slack is observed when frequency is over-boosted, translating into unnecessary energy losses. Then, the PID should slow down the processor allowing it to operate on lower frequencies. Another important aspect is how intervention of RAS mechanisms is perceived by the PID controller. A distinction should be made between functional correctness and performance variability. After a functional error is detected, the system's RAS mechanisms are responsible for correcting error and ensuring correct operation. The extra clock cycles of such mechanisms are translated into timing delay or, in the context of the controller monitored delay and timing noise, forcing the systems slack to fluctuate. It is important to note that the controller cannot guarantee correct functional operation but rather focuses in performance variability. To ensure that slack convergence to zero, the controller monitors slack at specific time intervals and decides to switch frequency and the associated supply voltage, choosing from a number of available operating points. In fact, in a spin-off experiment, we have verified that the number of DVFS points and therefore the granularity of the DVFS space is closely related to the settling time of slack, that is the time elapsed from the manifestation of a system disorder (in our case a RAS event) to the time at which the slack has entered and remained within a specified error band. Hence, based on the monitored slack and from the combination of the proportional, integral, and differential components, the new DVFS point is decided. The architecture of the PID controller has been elaborated in detail in [2] and is briefly shown in the next graph while the table summarizes what has been discussed so far.

After summarizing the basis of the PID concept, we will now focus on its implementation on top of a realistic embedded application which is the Spectrum

Sensing application. The core of the PID implementation is the monitor used to measure timing delay and estimate slack and the knobs used to switch to other DVFS points. For monitoring purposes, we use the C clock_gettime() function that allows measuring time intervals between code chunks with a millisecond accuracy. To switch to different operating points, we used a DVFS software driver that enables the effective exploitation of as many as 50 different points in the frequency range of 396–1050 MHz. The matching voltage levels have shown a linear relation between frequency and voltage. It should be stressed that the DVFS timing overhead is measured to be approximately 750 s. The main contribution to this is given by the time the protocol requires to set up a new voltage point (through I2C interface); this time can be reduced when using designs with on-chip regulators. Despite being able to use up to 50 DVFS points and have a fine-grain granularity of the DVFS space, we selected only 14 points for our implementation, since adding more frequency points does not reduce the settling time significantly. Specifically, the settling time is limited already to less than 2% with 14 DVFS points. This restriction of DVFS points is consistent with what is observed in state-of-the-art 22-nm processors, as measured on hosts running the Ubuntu® operating system, as shown in Fig. 7.6.

Lastly, after estimating the slack at a specific time interval ($s[n]$), the decision for the DVFS switch is based on the proportional, integral, and differential parts of the PID controller. The proportional component Kp produces an output value that is proportional to the current error value. The integral term Ki is proportional to

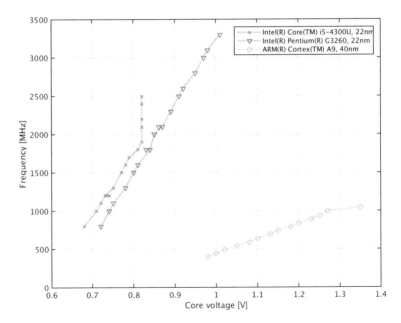

Fig. 7.6 Selected sensing dynamic voltage and frequency scaling (DVFS) points on the 40 nm ARM® core, compared with Intel® 22 nm processor cores

both the magnitude of the error and its duration therefore the integral part represents the memory of the controller. Lastly, the derivative part Kp is depending on the slope of the error. The gains of the three components are manually tuned during the experimentation process so as to achieve the minimum settling time. Tuning methods such as the Ziegler–Nichols can also be used, though not yet incorporated in our current implementation. With respect to computational complexity, the implemented scheme has a negligible impact. Specifically, the adjustment of speed and voltage happens at a rate of 100 Hz when the implementation is running real time. Per iteration six operations are needed for the PID controller, complemented with a search in an array of 14 elements. We can safely conclude that the computation load of the HARPA RTE stays below 10k instructions per second for a rate of 100 Hz.

7.2.4 GUI Support

The GUI of the sensing application runs on the laptop of the measurement setup. It allows to visualize the hardware and software status of the experiments with the HARPA RTE. Also, a check of the RAS signatures is done to detect functional errors, and if such an error is detected a rollback command is sent to the RAS code on the application platform. Additionally, the GUI implementation supports the following controls:

- An option to enable or disable the HARPA RTE PID controller
- An option to inject random errors
- An option to manually force an error
- An option to enable application logging for debugging the application

The GUI supports two tabs with graphical feedback as well as measurement data:

- A summary tab, shown in Fig. 7.7, keeping track of the following measurements over time: temperature, selected frequency/voltage combination, a buffer counter for the functional RAS mechanism, and a counter of the built-up processing slack. On the application platform, the slack is used for parametric mitigation in the PID controller.
- A detailed tab, shown in Fig. 7.8, keeping track of the following measurements over time: voltage, frequency, power, and energy consumption. The individual power measurements characterize the power consumption of the application processor, derived from a voltage measurement with the Arduino board.
- Measurement data on the present temperature, frequency, and voltage is updated continuously in the upper part of the GUI.

Since the application is not running real time, the measured wall clock time is translated to a real-time value, corresponding to the buffer interval duration. This is done to obtain a realistic view on the reaction speed of the HARPA RTE. The buffer

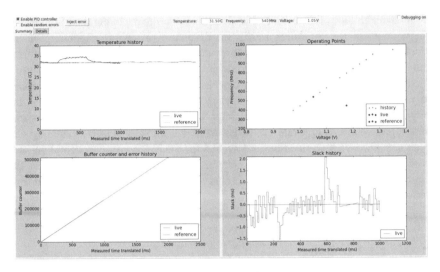

Fig. 7.7 Graphical user interface (GUI) in summary mode

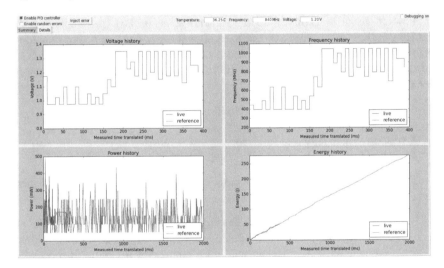

Fig. 7.8 GUI in detailed mode

count counts the number of completed buffer reads and prints a red dot if an incident occurred, requiring a RAS rollback. To compute the signatures, a reference run is done to compute reference signatures. During the reference run, the PID controller is disabled, and all measurement parameters are recorded as a reference. These reference values are plotted in green, while experimental data is plotted in blue. Such a reference run should be done before actual experiments.

7.3   Measurement Results

Three sets of experiments were performed on the measurement platform:

- Heat stress experiments to test if functional errors can be generated in an accelerated way, discussed in Sect. 7.3.1
- Reliability stress experiments with the HARPA RTE, shown in Sect. 7.3.2
- Workload stress experiments with the HARPA RTE, detailed in Sect. 7.3.3

7.3.1   Heat Stress Functional Error Rates

To measure the effect of heat stress on the experimental platform, the mean time before failure (MTBF) was monitored for an increasing temperature, and in function of frequency and voltage. At 40 nm, the CPU of the IMX6Q application board proved to be more robust to heat stress than expected. Specifically, heat alone was not sufficient to generate functional errors, and even at maximum frequency no errors were found. Only by increasing the voltage, we succeeded in generating measurable functional errors, and when they occurred these errors occurred frequently. So, this clearly substantiates the heavy overdimensioning of platforms fabricated in the current scaled technology nodes. We expect this overdimensioning to increase significantly for more advanced nodes.

Figure 7.9 shows the MTBF result for the following operating point:

- A temperature of 130 °C
- A clock speed of 996 MHz
- A voltage of 1.45 V

Two test cases were evaluated. In the first case, the application processor was kept idle for 5 s, and then spectrum sensing application code was run, together with the RAS monitor. At the laptop, the Mean Time before a failure was detected and logged. This was repeated during a time span of 5 h. In the second time, the experiment was repeated, but with an idle time of 220 s instead of 5 s. To smoothen the measurement data, a moving average filter was applied, resulting in the upper red curve for the second test case, and the lower blue curve for the first test case. This is as expected, since a lower idle interval results in a shorter recorded time before a failure. Hence, a lower processor load should result in less functional errors.

Since the applied heat and voltage stress was significant, the application board degraded after running a few test runs, resulting in irreparable functional errors. Two boards died while testing, and hence it was decided to emulate functional error injection for tests of the HARPA RTE.

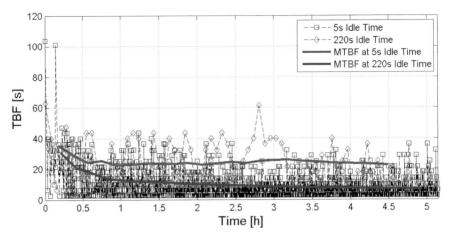

Fig. 7.9 Measured mean time before failure (MTBF) evolution over time

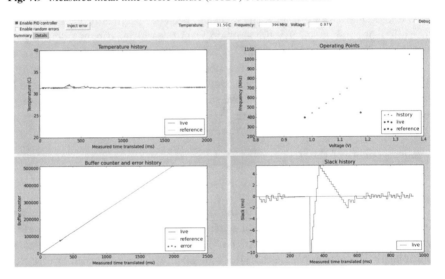

Fig. 7.10 Reliability stress response

7.3.2 Reliability Stress Compensation

To show the HARPA RTE reaction to reliability stress, a functional error is injected
while the spectrum sensing application is running, and the response is illustrated
in Fig. 7.10. In this test case, the base load is assumed to be running real time at
a clock speed of 400 MHz. Since the platform is not running real time, measured
time is translated to expected time at which the HARPA PID controller reacts in
real time. A consequence of this approach is that the PID controller overhead is not
accurately represented compared to the application load. As shown in Sect. 7.2.3,

the computational load of the PID controller is below 10k instructions per second, and hence negligible.

In the shown example, a functional error is triggered at 300 ms, and the resulting effect can be observed in the lower-right subplot, which represents the observed slack. As a consequence, the RAS mechanism detects an error and initiates a rollback. This results in performance variability, specifically the slack increases with 10 ms. This period corresponds to the spectrum sensing internal buffer memory, selected for a rollback, and is hence not arbitrary. To compensate the slack, the PID controller increases the clock, and subsequently the slack turns positive after 50 ms. Given the higher clock speed, the slack turns further positive, and the PID controller corrects this by lowering the clock speed back to its nominal setting to handle the workload.

With respect to processor temperature, a slight increase of temperature is observed when the PID controller reacts to the functional error, as shown in the upper-left subplot.

In summary, we observe that the PID controller of the HARPA RTE fully succeeds in compensating the performance variability, caused by reliability stress. So, the strict QoS requirements for the sensing application are fully met after the introduction of our parametric mitigation approach.

7.3.3 Workload Stress Compensation

To show the HARPA RTE reaction to workload stress, the application load is doubled on a single core, and the response is illustrated in Fig. 7.11. At 250 ms, the extra load is generated, and at 600 ms this load is removed. As a reaction, we see that the real-time performance is not met anymore, resulting in a negative slack. The HARPA RTE PID controller detects this and reacts by increasing the clock speed. When the extra load is removed, a positive slack is detected, and the PID controller relaxes the clock frequency.

With respect to processor temperature, we observe an increase from 32 to 35 °C during the period that the extra processor load is present.

In summary, we conclude that though the cause of the processor variability is different than in the previous section, the HARPA RTE can again effectively compensate it with the same mechanism.

7.4 Conclusion

To deal with performance variability, expected to increase in deep submicron technology nodes, the HARPA RTE was evaluated on the spectral sensing on the Freescale™ IMX6Q embedded platform. First, we showed that functional errors can be invoked by inducing temperature stress to the platform. We found

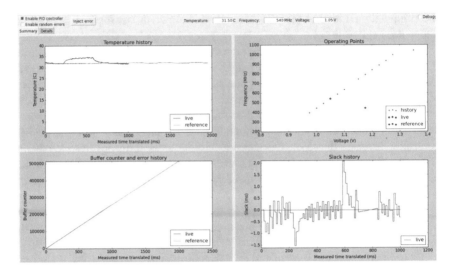

Fig. 7.11 Workload stress response

that for a 40-nm implementation the error behavior is abrupt, moving from the not detectable stage to frequent functional errors. This quickly results in non-correctable errors after a number of test runs. Hence, it was decided to opt for a software-emulated error rate to validate the HARPA RTE. It was shown that the HARPA RTE mechanism can effectively tackle both reliability and performance stress by adding a closed-loop PID controller that controls the DVFS settings of the embedded processor, and that reacts to performance variation. With respect to future studies, it is recommended to also cover communication between multiple processor cores, and specifically the effect on performance variation between different heterogeneous cores.

References

1. Mack, C. A. (2013). Keynote: Moore's Law 3.0. In *Proceedings of IEEE Workshop Microelectronics Devices* (p. 18).
2. Rodopoulos, D., Catthoor, F., & Soudris, D. (2015). Tackling performance variability due to RAS mechanisms with PID-controlled DVFS. https://doi.org/10.1109/LCA.2014.2385713
3. Stamoulis, D., Corbetta, S., Rodopoulos, D., Weckz, P., Debacker, P., Meyer, B. H., et al. (2016). Capturing true workload dependency of BTI-induced degradation in CPU components. In *GLSVLSI*. https://doi.org/10.1145/2902961.2902992
4. Rodopoulos, D., Zompakis, N., & Soudris, D. (2015). System scenarios HARPA RTE: Instantiation. HARPA deliverable 2.3.

Part IV
Knobs and Monitors

Chapter 8
Evaluating System-Level Monitors and Knobs on Real Hardware

Panagiota Nikolaou, Zacharias Hadjilambrou, Panayiotis Englezakis, Lorena Ndreu, Chrysostomos Nicopoulos, Yiannakis Sazeides, Antoni Portero, Radim Vavrik, and Vit Vondrak

8.1   Introduction

High-Performance Computing (HPC) infrastructures run a diverse set of applications with different Quality-of-Service (QoS) requirements [7]. These requirements come in various forms, such as operational power, performance, energy, cost, and availability. Naturally, HPC systems need to be configured in a way to satisfy those application requirements [15]. To configure a system, all the different system hardware and software knobs are set at specific settings, with many remaining at a default setting, and then, while the application runs, various monitors are observed to determine whether the various requirements are satisfied. Hardware and software knobs like Simultaneous Multithreading, DVFS, prefetching must be set to correct settings to achieve optimal energy and cost efficiency. Unsurprisingly, the configuration space is extremely large and such configuration search efforts are in practice ad hoc and nonoptimal.

One way to reduce the configuration search dimensionality and complexity is to reduce the requirements that need to be satisfied, as well as the monitors and knobs that need to be evaluated. Reduction of a problem's dimensionality is not a

P. Nikolaou (✉) · Z. Hadjilambrou · P. Englezakis · L. Ndreu · C. Nicopoulos · Y. Sazeides
University of Cyprus, Nicosia, Cyprus
e-mail: panagiota.nikolaou@cs.ucy.ac.cy; zhadji01@cs.ucy.ac.cy;
englezakis.panayiotis@ucy.ac.cy; lndreu01@cs.ucy.ac.cy; nicopoulos@ucy.ac.cy;
yanos@cs.ucy.ac.cy

A. Portero · R. Vavrik · V. Vondrak
VŠB - Technical University of Ostrava, IT4Innovations National Supercomputing Center,
Ostrava, Czech Republic
e-mail: antonio.portero@vsb.cz; radim.vavrik@vsb.cz; vit.vondrak@vsb.cz

© Springer International Publishing AG, part of Springer Nature 2019   167
W. Fornaciari, D. Soudris (eds.), *Harnessing Performance Variability in Embedded and High-performance Many/Multi-core Platforms*,
https://doi.org/10.1007/978-3-319-91962-1_8

new problem for computing system analysis [4, 9–11]. Such reduction, typically, relies on some form of statistical correlation, for example, principal component analysis [1, 17].

In this chapter, we explore the potential benefits of a methodology that would identify the minimum set of monitors and knobs needed to configure an HPC for optimal performance, availability, energy, and cost efficiency. To accomplish this, we use a correlation methodology that relies on data obtained from a detailed exploration of a configuration space. Even though a detailed exploration is what a practical search methodology should avoid, the analysis in this chapter is useful to assess the potential benefits that can come from a future methodology that reduces the search space.

Specifically, for this investigation, we use Floreon+ application, a flood prediction and management application, running on an HPC platform. The analysis considers data obtained using eleven system monitors when exploring many settings for six knobs. This analysis is performed for individual and combinations of the following requirements: performance, power, availability, energy, and cost. The results reveal that the configuration space can be reduced considerably. Additionally, the results show non-obvious correlations between requirements, monitors, and knobs. This motivates the need for further research to determine a practical configuration space reduction methodology for HPC systems. Additionally, with the exploration of the potential benefits of methodology that would identify the minimum set of monitors and knobs needed to configure an HPC system, we provided an exhaustive characterization of all the monitors and knobs. The characterization reveals the following insights: (a) Floreon+ is CPU bound (not memory intensive) and the performance scales linearly with CPU frequency and (b) Floreon+ performance requirements are satisfied with a CPU frequency lower than the servers nominal CPU frequency. This implies that there is a power headroom that can be exploited for dealing with performance variability caused by faulty hardware.

The rest of the chapter is organized as follows: Sect. 8.2 covers background related to the Floreon+ [23] application and HPC organization. Section 8.3 describes the characterization and experimental framework and correlation analysis, and Sect. 8.4 presents and discusses experimental results. This chapter concludes in Sect. 8.5.

8.2   Background

8.2.1   HPC Application (Floreon+)

We use Floreon+ (FLOod REcognition On the Net) [23], an HPC application with high QoS requirements. Floreon+ is an online system for monitoring, modeling, prediction, and support disaster flood management [18]. The system focuses on acquiring and analyzing relevant data in near-real time. The data are used to provide short-term flood prediction by running hydrologic simulations.

The main processes of Floreon+ application are organized as follows:

1. Get information about actual river and reservoir situation.
2. Rainfall-runoff (RR) modeling: simulation of surface runoff.
3. Hydrodynamic (HD) modeling: flood lake simulations, flood maps, simulations of water elevation and water velocity, a real-time hydrological model for flood prediction, water quality analysis, etc.
4. Erosion modeling: simulation of water erosion.
5. Collection and archiving of flood data that can be used to estimate the magnitude of a flood based on historical evidence.

In this work, we are investigating the uncertainty of the Rainfall-runoff (RR) modeling which is the most computationally intensive part [18, 23].

The application framework for the uncertainty of RR model provides an environment for running multiple simulations every repetition, when new data arrives on an HPC system. The uncertainty contains information about how accurate is the solution that RR model provides. RR model is a dynamic mathematical model, which transforms rainfall to flow at the catchment outlet.

The uncertainty is computed as Monte Carlo samples. The Monte Carlo method gives a straightforward way of massive parallelism by increasing the number of random values working concurrently to obtain numerical results. Previous experiments [18] exhibit a good scalability of the Monte Carlo method in an HPC cluster with 64 nodes of 16 cores each. Consequently, the proposed methodology may be appropriated in any other HPC framework where the Monte Carlo method is employed.

Figure 8.1 shows the normal operation of Floreon+. As Fig. 8.1 shows, a batch of Monte Carlo iterations is running in a number of nodes (Server 1–Server n) in such a way that application's QoS requirements are satisfied. Each interval indicates the execution of a different simulation. For example, the first interval refers to the first simulation. After the execution of all the Monte Carlo iterations, the results send to a master server for processing. The total simulation time includes the execution time of Monte Carlo iterations and the time needed to process the results. When a simulation ends, the servers remain idle for a set of period. The duration of this period is determined by the availability of the new batch of data. Under normal operation (fault-free), the simulation always finishes within the time constraint. There are some cases where a fault on a component can delay the execution of the simulation, as shown in Fig. 8.2. These cases can be categorized in the following:

1. Delay the execution of the simulation but still the simulation finishes within the time constraint. The availability of the system does not decrease.
2. Delay the execution and violate the timing constraint with the same number of servers. Thus the results of this simulation are useless and the availability of the system decreases.

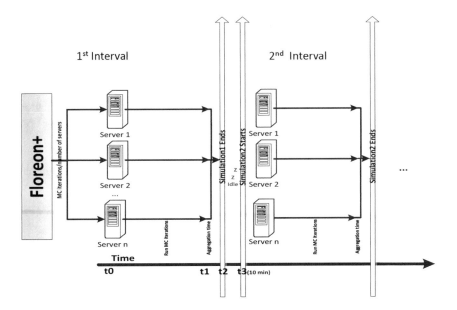

Fig. 8.1 Floreon+ normal operation

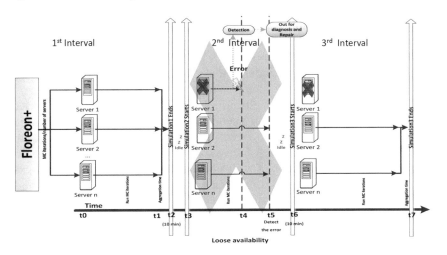

Fig. 8.2 Floreon+ operation with faults

3. Delay the execution and violate the timing constraint with less servers. In this
   case, the faulty server needs to be taken offline until it is repaired or replaced. In
   this case, the results are lost and the availability decreases.

Figure 8.2 illustrates the last case where the server needs to be taken offline until it is repaired or replaced. As shown in the figure, the number of servers in the third interval is decreased and the job is assigned to the remaining servers until the faulty server is repaired or replaced.

Next the various requirements presented for this application.

8.2.1.1 Application Requirements

Floreon+ has two running operation modes, the standard operation mode and the emergency operation mode. Both have different requirements. Standard operation mode is the default operation of the system. In this operation, the weather is favorable and the flood warning level is below the critical threshold.

On the other hand, on the emergency operation mode the water in the rivers rises due to continuous rain or free-flowing streams that are created due to heavy rainfall on small areas. During this operation mode much more accurate and frequent simulation computations are needed and the results should be provided as soon as possible. In this work, we focus on the emergency operation which has tighter timing requirements.

8.2.1.2 Availability and Quality of Service Requirements

The reliability and availability target of Floreon+ running on emergency operation is accomplished through a combination of hardware/software mechanisms and policies. This also aims at satisfying the QoS requirements even in the presence of errors and offline servers. These mechanisms typically rely on hardware and software **monitors** and **knobs**.

In general, when a server fails and if its repair time is expected to be long, the system software migrates the failed job to another server. The failure of the server is detected by a hardware or software monitor. The migration is possible because of QoS and availability reasons. HPC systems are over-provisioned with spares for dealing with errors and offline servers. Server over-provisioning is determined by the availability of a system. The less available system, the more servers needed [22].

Because of its significance, emergency operation requires responsiveness in 10 min for each simulation. Also it must provide high levels of availability, two nines (0.99), which may require over-provisioning with extra servers to deal with various hardware failures.

Floreon+ and other offline services can run together (collocated) to improve utilization [21, 31]. Specifically, when Floreon+ satisfies the QoS requirements without using all the available cores in a server, the remaining cores can run other services. This must be done without affecting the QoS of Floreon+ and violating its requirements. Since we are going to explore the emergency operation, Floreon+ is running in isolation, utilizing all the available resources without any other service concurrently runs on the same server.

Table 8.1 Floreon+
requirements

Performance	Simulation $\leq 10\,\text{min}$
Accuracy	$\geq 10^4$ MC iterations
Availability	≥ 0.99
Power	$\leq 81\,\text{W}$
Energy	$\leq 48{,}600\,\text{J}$

8.2.1.3   Accuracy on Monte Carlo Iterations

It is of utmost importance that the results are as precise as they can. The precision
of the simulated results is based on the number of Monte Carlo iterations [24]. It
has been shown in [24] that the number of iterations has to be in the order of 10^4 to
10^5 to obtain a satisfying precision. In this work, we assume 20,000 Monte Carlo
iterations that have to be computed before the deadline (i.e. 10 min, since new input
data arrives from weather stations) as the baseline configuration.

Table 8.1 summarizes the specific values for the requirements of the Floreon+
application.

8.2.2   Available Monitors and Knobs in HPC Systems

The incessant shrinking of technology feature sizes comes with a whole new set
of challenges for CPU designers. The diminutive transistors are increasingly more
prone to process variability accelerated aging artifacts. In fact, the lifetime of near-
future chips is expected to be affected, as new technologies exacerbate reliability
issues. As a result, the nature of chip design has changed, since it has to tackle a lot
of probabilistic phenomena. This shift in design practices has forced the designers to
spend additional resources, in order to maintain functional correctness and Quality-
of-Service (QoS) requirements (e.g., energy/power, performance, etc.).

There is an abundance of phenomena that cause the lifetime to shrink and
introduce variability to the behavior of a CPU. Some of them are present at time
zero, i.e., right after a chip is manufactured. This is known as *time-zero variability*.
Time-zero variability is created by the chip manufacturing process itself. This
can be attributed to phenomena like random dopant fluctuations and line-edge
roughness, among others. On the other hand, the chip's behavior also changes
during its lifetime. This is known as *time-dependent variability*. Time-dependent
variability is intertwined with the operating conditions of the chip, like the workload,
temperature, frequency, and voltage. The effects of both variability types on a chip
can be estimated by designers, in an attempt to mitigate their impact.

To deal with the effects of variability, CPU designers introduce guard bands on
voltage and frequency, based on accurate models. Guard bands make sure that—
irrespective of the severity of the variability (time-zero or time-dependent)—the
chip's correctness and QoS behavior will be maintained.

Since errors and/or QoS violations may now occur in the system, the use of system-wide, strategically located *monitors and knobs* becomes mandatory. *Monitors* supervise the system and make sure that metrics such as QoS are satisfied. If a metric is violated, then the *knobs* take action to compensate for the deviation. *Knobs* are, essentially, actuators that try to restore the nominal value of a violated metric. There are two approaches in the use of monitors and knobs. They can be used *proactively*, or *reactively*. In the first case, the monitors observe the evolution of certain operating parameters. They try to mitigate any deviation n in these parameters, in order to delay (as much as possible) any violation. In the second case, the monitors and knobs wait for a violation to happen, and then they take action to correct the deviation. Monitors and knobs can be implemented in hardware and/or software.

A large number of monitors and knobs exist in HPC systems. Monitors and knobs in real systems can be categorized into the following categories depending on what metric they influence: performance, power, temperature, and reliability.

Table 8.2 shows the requirements, monitors, and knobs that are going to be explored in this work. This subset is by no means comprehensive and future work should consider a larger set.

Monitors that are going to be investigated are execution time, Instructions per Cycle (IPC), Misses per kilo Instructions (MPKI), DRAM, CPU and peak power, and CPU temperature. Also, Mean Time Between Failures (MTBF) per server and for the whole system is used through analytical models and Failures In Time (FIT) rates [25]. Finally, capital (CAPEX) and operational (OPEX) expenses are going to be estimated based on publicly available info, e.g., list prices, and runtime measurements. CAPEX expenses include infrastructure, server and networking equipment costs, whereas OPEX expenses include power and maintenance costs.

Table 8.2 also shows the different knobs that we are going to experiment with. As the table shows, we use Simultaneous Multi-Threading (SMT) [28], Dynamic Voltage and Frequency Scaling (DVFS) [30], data prefetchers [27], and Intel's Turbo Mode [14]. Also, this work provides results with and without redundant cores.

Table 8.2 Application requirements, monitors, and knobs

Requirements	Monitors	Knobs
Performance	Execution time	DVFS
Power	IPC	SMT
Energy	DRAM power	DRAM protection
Availability	CPU power	Turbo mode
Cost	Peak power	Prefetchers
	CPU temperature	Redundancy
	MPKI	
	Server MTBF	
	System MTBF	
	Capex expenses	
	Opex expenses	

Redundant cores are used to improve the reliability of the system by migrating the running thread of a faulty core to a spare [13]. For the redundancy scenario, it is assumed that half of the cores remain idle to provide higher availability. On the other hand, in the scenario without redundancy all the available physical resources are utilized. Furthermore, this study explores the implication of using two different DRAM Protection Techniques (No Protection or ChipkillDC).

DRAM is protected from errors by using extra devices per DIMM to store Error Correction Code (ECC) bits. Modern processors usually support Chipkill with 16 ECC bits to protect 128 data bits that are interleaved across two DIMMs placed in two channels [3, 5, 6]. This is referred to as ChipkillDC or Lockstep where it can correct all the errors in a single device and detects all the errors in two devices [5]. Chipkill can waste bandwidth, hurt performance, and increase energy consumption [2, 16, 29]. On the other hand, No Protection does not provide any protection on DRAM.

8.3 Experimental Framework and Correlation Analysis

8.3.1 Experimental Framework

For evaluation, we use an HPC cluster with dual socket Intel Xeon E5-2640 v3 system configuration, as shown in Table 8.3. We run each experiment five times, and each time we observed execution time, instructions per cycle (IPC), misses per kilo instructions (MPKI), CPU, DRAM power, and CPU temperature. The results presented are calculated by removing minimum and maximum values and calculating the average.

Prefetchers and DRAM protection techniques changed by accessing BIOS, through a BIOS Serial Command Console interface (CLI) [12].

Our evaluation used *Floreon+*, an HPC application with a dataset of 44KB (Tiny dataset). This is a representative dataset size for the application purposes because it uses 5 days observations to provide predictions for the next 2 days. All the power numbers are collected using the Likwid-powermeter [19] which allows monitoring the power consumption of CPU and DRAM at any given time. The results are used to calculate total power and peak power numbers.

Table 8.3 Server configuration

Number of CPUs	2
CPU	Intel Xeon E5-2640 v3
Number of cores per CPU	8
Number of threads per core	2
Channels per CPU	4
DIMMs/channel	2
DIMM capacity	16 GB

Table 8.4 Values of knobs

Knobs	Value
DVFS	1.2, 1.7, 2.2, 2.6 (GHz)
SMT	Disable, enable
DRAM protection	No protection, ChikillDC
Turbo mode	Disable, enable
Prefetchers	Disable, enable
Redundancy	Disable, enable

To track CPU temperature, we use lm-sensors [20]. To estimate Server MTBF and System MTBF monitoring values, as well as availability values, we use different analytical models based on binomial probabilities.

Also, to estimate CAPEX, OPEX expenses as well as total cost, we use COST-ET and AMPRA tools proposed in [8, 22].

For each experiment, an initial population of 2 server modules is used. The number f servers can change depending on availability and performance requirements.

For a baseline configuration, we select the one that is currently used to run this application and includes the following values for each parameter: SMT: OFF, Frequency: 2.6 GHz, DRAM Protection: No Protection, Turbo Mode: Enable, Redundancy: 0 (No), Prefetchers: ON.

The data used for correlation analysis are obtained by exploring the 128 combinations of knobs presented in Table 8.4. For each configuration combination, eleven monitor values are recorded. These eleven values are then used to determine the values of five metrics each corresponding to a specific requirement.

The values of the requirements for a given configuration are obtained as follows: performance from execution time, power from CPU and DRAM power, energy product of execution time and power, availability from server MTBF, server MTTR, and number of servers, and cost using monitors for power, performance, availability using the COST-ET tool [8]. The total data set used for this analysis is, therefore, a $128 \times 11 \times 5$ data matrix, where 128 are the combinations of knobs, 11 are the monitors values, and 5 are the requirements.

In this section, we describe the methodology used for correlation analysis to reduce the number of requirements, monitors, and knobs used to configure an HPC system. The data that drive this analysis are obtained as discussed in Sect. 8.3.1. The value obtained for each requirement for a given experiment is normalized to give each requirement equal weight. The normalization is done by computing the mean and standard deviation for each requirement and then by subtracting the mean and dividing the standard deviation by each value of requirements as in [10]. A configuration is considered to be successful if it satisfies each individual application requirement. To rank successful configurations, we determine an overall score for a configuration by multiplying each normalized requirement value with an equal weight. In this case, since we have five requirements, we multiply each value with 0.2. The correlation analysis is done using the R statistical language [26].

8.3.1.1 Heuristic Correlation Analysis

This analysis explores the correlation between requirements, monitors, and knobs. The methodology used is as follows:

1. For a given requirement, we compute the correlation coefficient (using Pearson correlation analysis) with all the other requirements. For each pair (x_i, x_j) of requirements i and j, where $i \neq j$, that exhibit significant correlation coefficient (above a 90% threshold[1]), we check which of the two requirements can be removed. The requirement that shows smaller correlation coefficient with all the other requirements is removed from the list. This process is iterated over all remaining requirements. The same process is repeated for the monitors.
2. For the remaining requirements and monitors, we compute the correlation coefficient between each requirement and all remaining monitors and select the monitor with the highest correlation.
3. For all the remaining monitors and all the available knobs, we compute the correlation coefficient between them and knobs that have a correlation coefficient above a 40% threshold (see footnote 1) are kept.

This correlation analysis aims to reduce the number of configurations that need to be explored to determine the configuration that provides the highest satisfaction of the requirements according to the ranking in Sect. 8.3.1. Specifically, the analysis returns a subset of the monitors and knobs. All possible configuration combinations are then evaluated for the selected knobs. For the knobs that are not selected, we used the baseline configuration values (we have confirmed that it does not really matter which value you use for them). Afterward, the selected configurations are sorted according to the selected monitor(s) value(s) (if there is more than one monitor, equal weighting is used to combine them). The top ranked configuration using the selected knobs and monitors is then compared with the configuration that considers all. Their difference is measured as the maximum negative % difference for any of the requirements (if it is 0 for all, it means it matches the best possible configuration).

8.4 Results

8.4.1 Implications of Different Combinations of Knobs in Performance

Before showing the results of the proposed methodology, we show the results of different analysis that we provided.

The first case study investigates how performance is affected by each combination of knobs. For this analysis, we are using as metrics for the performance the

---

[1]The specific thresholds are picked on empirical analysis not shown here.

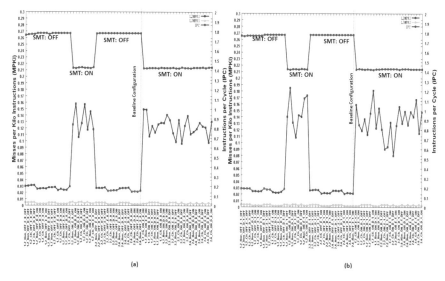

Fig. 8.3 Misses per Kilo Instructions (MPKI) for L3 and L2 cache memories and Instructions per Cycle (IPC) (**a**) without redundant cores (**b**) with redundant cores

Misses per kilo Instructions (MPKI) for L3 cache and L2 cache, the Instructions per Cycle (IPC), and the execution time. In the end, we are going to isolate which of the three are representative for this application. The results of the MPKI and IPC are shown in Fig. 8.3.

The x-axis presents different combinations of knobs while the $y1$-axis (left) shows the MPKI values and the $y2$-axis (right) shows the IPC values. Each combination of the knobs is encoded with the following structure:

[frequency_DRAMProtection_SMT_TurboMode_Redundancy_Prefetchers]

Figure 8.3a shows how the different knobs affect the IPC and MPKI without any redundant cores while Fig. 8.3b shows how the different knobs affect the IPC and MPKI including redundant cores. The redundant cores are used as spares to increase the availability of the system in case of a failure. The dotted line shows the configuration currently used in the servers.

In both cases (Fig. 8.3a, b), no meaningful results can be extracted about the correlation of different combinations of knobs with performance. The only outcome concerns SMT. In the case SMT is ON, the IPC decreases and MPKI increases due to L2 memory coherency of the extra thread running on the same core. On the other hand, when SMT is OFF the IPC increases and the L2 MPKI shows the reversed behavior. Since the application is not memory intensive, L3 MPKI is very low.

The results clearly suggest that IPC and MPKI are not the representative metrics for this application since, except SMT, they are not affected by a combination of knobs.

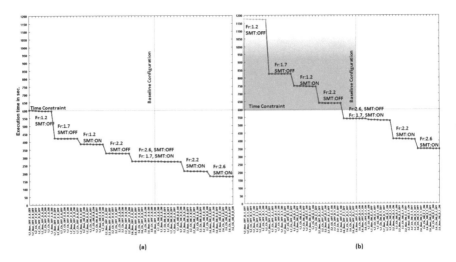

Fig. 8.4 Execution time results for the different combinations of knobs (**a**) without redundant cores (**b**) with redundant cores

In Fig. 8.4a, b, we show the sorted execution time in seconds for the different combinations of knobs. The time constraint was set to 600 s (10 min), as discussed earlier.

As shown in both figures, different combinations of knobs have different effects on the execution time. The steps shown in Fig. 8.4 show a trend. The labels above of each step show the knobs that remain unchanged for that plateau. We observe the same trend in the results with and without redundant cores. The main difference is that Fig. 8.4b (with redundant cores) is moved higher in the y-axis.

Our analysis shows that without redundancy all the combinations of knobs satisfy the time constraint. At the same time, this does not happen when the same experiments are run with redundant cores. Moreover, there are a lot of scenarios that can provide better execution time than the baseline in each configuration.

Next, we explore the trade-off between power, reliability, and cost using only the combination of knobs that do not violate the time constraint in both cases (with and without redundancy).

8.4.2 Implications of Different Combinations of Knobs on Reliability

This case study investigates how the Mean Time Between Failures (MTBF) is affected by each combination of knobs that do not violate the time constraint.

Figure 8.5a, b show the effects of different combinations of knobs in the system's MTBF without and with redundancy, respectively. The y-axis represents

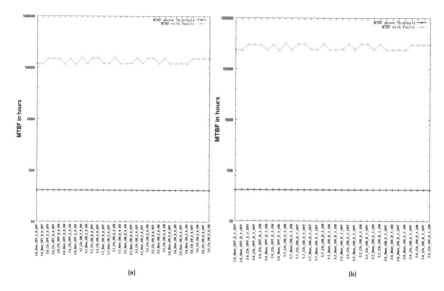

Fig. 8.5 Mean Time Between Failures (MTBF) on a logarithmic scale for each combination of knobs (**a**) without redundant cores (**b**) with redundant cores

the system's MTBF on a logarithmic scale. The target reliability in this experiment is shown with the red line, whereas the calculated reliability is shown with green line.

As shown in both cases (Fig. 8.5a, b), estimated MTBF is above the threshold. It is clear though that the configuration with the redundant cores provides better reliability.

These results can be changed in a larger server farm, where MTBF becomes lower. Especially, without redundancy (in cores) it can become even lower than the target MTBF. Furthermore, having redundant, idle resources can provide higher reliability because it is less likely for errors to manifest in idle resources. On the other hand, it is very important to have as much resources as needed but not increase the cost estimation. The above analysis indicates that the combination of knobs with redundant cores provides better resiliency.

8.4.3 Implications of Different Combinations of Knobs on Power

The peak and average power for CPU and DRAM per combination of knobs is presented in Fig. 8.6a, b. The peak (pink line) and average (bars) power results are presented in watts on the primary $y1$-axis. The execution time is presented on the secondary axis ($y2$-axis). Shown in the figure, peak power is similar for all the

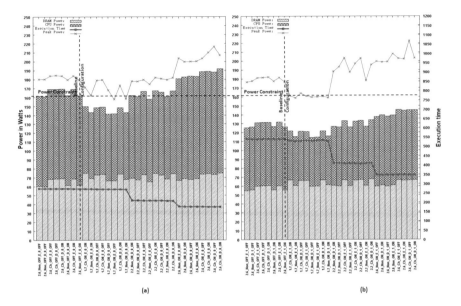

Fig. 8.6 Power results for each combination of knobs (**a**) without redundant cores (**b**) with redundant cores

scenarios (with and without redundancy). Also, taking into account only peak power will result in suboptimal observations. On the other hand, average power results (bars) show that the power consumption on the scenario with redundancy is lower that the scenario without redundancy. This is reasonable due to the cores that are remaining idle in the scenario with redundancy to improve the reliability of the system.

The combinations of knobs selected for the following case studies have power consumption less than the baseline configuration in both scenarios (with and without redundancy).

8.4.4 Implications of Different Combinations of Knobs on Energy and Cost

Now we are going to investigate how the different combinations of knobs perform, using monitors that encapsulate more than one metric. Firstly, we are going to explore the product of Power and Delay (Energy) for those combinations of knobs that satisfy the constraints of performance and power. In this case, we are going to find which combination of knobs can provide the best configuration to improve energy and cost.

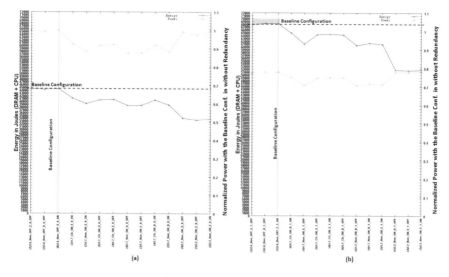

Fig. 8.7 Energy (Power X Delay in Joules) results for each combination of knobs (**a**) without redundant cores (**b**) with redundant cores

Figure 8.7 presents the Energy results of the different combinations of knobs. As we can see from the figure, there is one combination of knobs in the scenario with redundant cores (highlighted in red) that consumes more energy than the baseline. This combination of knobs was removed from both scenarios (with and without redundancy). The exploration for the cost continued with the remaining combination of knobs.

Figure 8.8a, b present the TCO of the different combinations of knobs and the number of total servers that are needed for each combination to match the performance of the baseline configuration in each scenario (with and without redundancy). TCO results include all the parameters (power, performance, temperature, reliability, etc.) that are investigated in this analysis.

An important observation from Fig. 8.8a, b is that infrastructure cost and network cost have a negligible impact on the total cost, whereas power, maintenance, and server costs consume a significant portion of the total cost. Finally, from the Figure we can conclude that the best scenarios in both cases can reach TCO savings in the range of 19–30%.

Overall, the best configuration in terms of all the metrics that we investigated in this analysis to run Floreon+ for the best (without redundancy) and the worst scenarios (with redundancy) is the following:

1. SMT:ON
2. Frequency: 2.2 GHz
3. DRAM Protection: Non

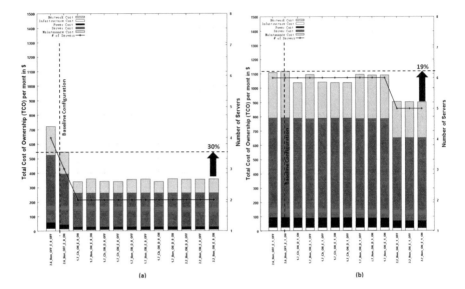

Fig. 8.8 Cost results for each combination of knobs (**a**) without redundant cores (**b**) with redundant cores

Table 8.5 Correlation between monitors and requirements

Requirements	Monitors
Performance	Execution time
Availability	Capex cost
Power	CPU power
Cost	Capex cost
All	Execution time, CPU power, Capex cost

Intel's Turbo mode and cache prefetchers can be either enabled or disabled. Based on our findings, these two parameters do not seem to affect any of the metrics monitored in this analysis. The findings of this analysis call for IT4I owners and researchers to select the above configuration that maximizes the TCO savings.

8.4.5 Results of the Heuristic Correlation Analysis

Our analysis reveals that energy is strictly correlated with performance, and thus we removed energy from the list of requirements and the presented results.

Table 8.5 shows the results of correlation analysis between requirements and monitors.

As we can see from the Table, Performance requirement can be monitored by Execution Time. Availability can be monitored by Capex expenses. This correlation is not obvious because Capex cost has an indirect relation with Availability. The

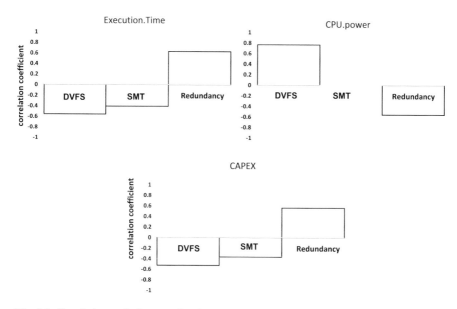

Fig. 8.9 Correlation analysis presenting the correlation coefficient between monitors and knobs

correlation of Capex expenses and availability comes from the extra servers that are needed for over-provisioning. The more the servers needed for over-provisioning, the more Capex expenses are incurred. Moreover, the more the servers needed for over-provisioning, the less the System MTBF due to the higher probability of errors. So, this leads to a smaller availability for this system. As the table, also, shows, CPU power is the appropriate metric for the total power. This happens because Floreon+ is not a memory-intensive application, so it has a negligible impact on DRAM power. For the total cost and overall combination of requirements, Capex is the correlated monitoring value.

Finally, Fig. 8.9 shows the correlation between knobs and monitors. X-axis presents the remaining list of knobs for each monitor, whereas the y-axis presents the correlation coefficient between knobs and monitors. The correlation coefficient ranges from -1 to $+1$. A value closer to $+1$ means that this knob has almost a linear relation with the monitor. A value closer to -1 means that this knob has an inverse relation with the monitor. A value closer to 0 means that there is no correlation between the knob and the monitor.

As we can see from the figure, DVFS, SMT, and Redundancy are the selected knobs. On the other hand, DRAM Protection, Turbo Mode, and Data prefetching can be reduced from the search space. This is due to the low cache and low memory pressure of the Floreon+ application (L3MPKI: 0.0034, L2MPKI: 0.013).

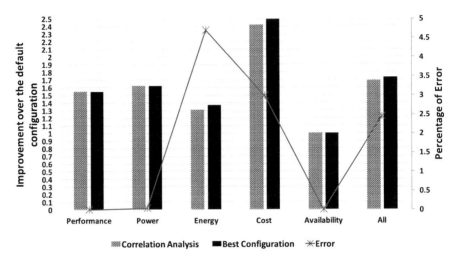

Fig. 8.10 Improvement between the default configuration, correlation analysis predicted configuration, and the best configuration for different requirements

8.4.6   Validation of Heuristic Methodology

Figure 8.10 validates the heuristic analysis by evaluating the best configuration for each requirement using the subset of monitors and knobs revealed from the correlation analysis.

In this graph, we show the normalized with the default configuration, improvement for each individual requirement and all requirements between the best configuration using the selected monitors and knobs and the best configuration using all monitors and knobs. As can be seen from Fig. 8.10, all the requirements can be improved by changing the system configuration. For example, performance can be improved by 1.5× and cost by 2.4× related to the default configuration. Also, the combination of all the requirements can be improved by 1.7×. Moreover, the figure shows that the configurations based on the correlation analysis are very close to the results with the best configuration. We measure the error between the two configurations, and we found a maximum value of 5%.

8.5   Conclusions and Future Work

This chapter describes an oracle methodology that investigates combinations of different knobs and the effects they have on the various monitored values. This aims to explore the potential benefits from a yet to be determined, realistic methodology that can reduce the number of monitors and knobs that configure an HPC system that runs a specific application while satisfying the application requirements. To this end, this analysis reduces an eleven space of system monitors to three and six-system space knobs to three. Also, a characterization analysis of the application reveals that Floreon+ performance requirements are satisfied with a CPU frequency lower than the servers nominal CPU frequency.

Findings of this work motivate future research. One direction is to find a realistic methodology probably based on correlation analysis that will be generic for many applications. Another direction is to minimize the data sample used.

References

1. Abdi, H., & Williams, L. J. (2010). Principal component analysis. *Wiley Interdisciplinary Reviews: Computational Statistics, 2(4)*, 433–459.
2. Ahn, J. H., Jouppi, N. P., Kozyrakis, C., Leverich, J., & Schreiber, R. S. (2009). Future scaling of processor-memory interfaces. In *Conference on High Performance Computing Networking, Storage and Analysis* (pp. 42:1–42:12).
3. Ankireddi, S., & Chen, T. (2014, February). Configuring and using DDR3 memory with HP ProLiant Gen8 Servers, Best Practice Guidelines for ProLiant servers with Intel Xeon processors.
4. Annavaram, M., Rakvic, R., Polito, M., Bouguet, J.-Y., Hankins, R. A., & Davies, B. (2004). The fuzzy correlation between code and performance predictability. In *Proceedings of the 37th annual IEEE/ACM International Symposium on Microarchitecture* (pp. 93–104). New York: IEEE Computer Society.
5. BIOS and Kernel Developers Guide (BKDG) for AMD Family 10h. (2010, April).
6. BIOS and Kernel Developers Guide (BKDG) for AMD Family 15h. February (2014).
7. Brewer, E. (2001). Lessons from giant-scale services. *Internet Computing, 5(4)*, 46–55.
8. Hardy, D., Kleanthous, M., Sideris, I., Saidi, A., Ozer, E., & Sazeides, Y. (2013). An analytical framework for estimating TCO and exploring data center design space. In *International Symposium on Performance Analysis of Systems and Software* (pp. 54–63).
9. Hoste, K., & Eeckhout, L. (2006). Comparing benchmarks using key microarchitecture-independent characteristics. In *2006 IEEE International Symposium on Workload Characterization* (pp. 83–92). New York: IEEE.
10. Hoste, K., & Eeckhout, L. (2007). Microarchitecture-independent workload characterization. *IEEE Micro, 27*(3), 63–72.
11. Hoste, K., Phansalkar, A., Eeckhout, L., Georges, A., John, L. K., & De Bosschere, K. (2006). Performance prediction based on inherent program similarity. In *Proceedings of the 15th International Conference on Parallel Architectures and Compilation Techniques* (pp. 114–122). New York: ACM.
12. Hp rom-based setup utility user guide. February HP TR (2014).
13. Huang, L., & Xu, Q. (2010). Characterizing the lifetime reliability of many core processors with core-level redundancy. In *2010 IEEE/ACM International Conference on Computer-Aided Design (ICCAD)* (pp. 680–685). New York: IEEE.
14. Intel. (2008). *Intel Turbo Boost Technology in Intel Core microarchitecture (Nehalem) Based Processors*. White paper. Technical report, November 2008.
15. Iordache, A., Buyukkaya, E., & Pierre, G. (2015). Heterogeneous resource selection for arbitrary HPC applications in the cloud. In *IFIP International Conference on Distributed Applications and Interoperable Systems* (pp. 108–123). Berlin: Springer.
16. Jian, X., Duwe, H., Sartori, J., Sridharan, V., & Kumar, R. (2013). Low-power, low-storage-overhead Chipkill correct via multi-line error correction. In *International Conference on High Performance Computing, Networking, Storage and Analysis* (pp. 24:1–24:12).

17. Jolliffe, I. (2002). *Principal component analysis*. Wiley Online Library.
18. Kuchar, S., Golasowski, M., Vavrik, R., Podhoranyi, M., Sir, B., & Martinovic, J. (2015). Using high performance computing for online flood monitoring and prediction. *International Journal of Environmental, Ecological, Geological and Geophysical Engineering, 9*(5), 267–272.
19. likwid-powermeter. Tool for accessing RAPL counters on Intel processor. https://github.com/RRZE-HPC/likwid/wiki/Likwid-Powermeter
20. lm-sensors. http://www.lm-sensors.org/wiki/man/sensors-detect
21. Mars, J., Tang, L., Hundt, R., Skadron, K., & Soffa, M. L. (2011). Bubble-up: Increasing utilization in modern warehouse scale computers via sensible co-locations. In *44th Annual International Symposium on Microarchitecture* (pp. 248–259).
22. Nikolaou, P., Sazeides, Y., Ndreu, L., & Kleanthous, M. (2015). Modeling the implications of dram failures and protection techniques on datacenter TCO. In *Proceedings of the 48th International Symposium on Microarchitecture, MICRO-48* (pp. 572–584). New York: ACM.
23. Portero, A., Kuchař, Š., Vavřík, R., Golasowski, M., & Vondrá, V. (2014). System and application scenarios for disaster management processes, the rainfall-runoff model case study. In *IFIP International Conference on Computer Information Systems and Industrial Management* (pp. 315–326). Berlin: Springer.
24. Portero, A., Vavrik, R., Kuchar, S., Golasowski, M., Vondrak, V., Libutti, S., et al. (2015). Flood prediction model simulation with heterogeneous trade-offs in high performance computing framework. In *29th EUROPEAN Conference on Modelling and Simulation ECMS*.
25. Sridharan, V., Stearley, J., DeBardeleben, N., Blanchard, S., & Gurumurthi, S. (2013). Feng Shui of supercomputer memory: Positional effects in dram and SRAM faults. In *International Conference on High Performance Computing, Networking, Storage and Analysis* (pp. 22:1–22:11).
26. Team, R. C. et al. (2013). *R: A language and environment for statistical computing*, Citeseer.
27. Tendler, J. M., Dodson, J. S., Fields, J., Le, H., & Sinharoy, B. (2002). Power4 system microarchitecture. *IBM Journal of Research and Development, 46*(1), 5–25.
28. Tullsen, D. M., Eggers, S. J., & Levy, H. M. (1995). Simultaneous multithreading: Maximizing on-chip parallelism. *SIGARCH Computer Architecture News, 23*(2), 392–403.
29. Udipi, A. N., Muralimanohar, N., Balsubramanian, R., Davis, A., & Jouppi, N. P. (2012). LOT-ECC: Localized and tiered reliability mechanisms for commodity memory systems. In *39th Annual International Symposium on Computer Architecture* (pp. 285–296).
30. Weiser, M., Welch, B., Demers, A., & Shenker, S. (1994). Scheduling for reduced cpu energy. In *Proceedings of the 1st USENIX Conference on Operating Systems Design and Implementation, OSDI '94*. Berkeley: USENIX Association.
31. Yang, H., Breslow, A., Mars, J., & Tang, L. (2013). Bubble-flux: Precise online QoS management for increased utilization in warehouse scale computers. In *40th Annual International Symposium on Computer Architecture* (pp. 607–618).

Chapter 9
Monitor and Knob Techniques in Network-on-Chip Architectures

**Davide Zoni, Panayiotis Englezakis, Kypros Chrysanthou, Andrea Canidio,
Andreas Prodromou, Andreas Panteli, Chrysostomos Nicopoulos,
Giorgos Dimitrakopoulos, Yiannakis Sazeides, and William Fornaciari**

9.1 Introduction

The everlasting demand for more processing power has rendered the multi-/many-core paradigm the de facto implementation choice in modern CPU architectures. Specifically, multi-core designs are now employed in both embedded and high-performance general-purpose systems, due to their flexibility, performance prowess, and fine-grain control over the chip's power envelope. At the same time, multi-core systems have accentuated the criticality of the on-chip interconnect backbone; traditional bus-based solutions suffer from limited bandwidth capabilities, which cannot cope with increasing numbers of processing cores. Hence, Networks-on-Chip (NoCs) have emerged as the standard interconnect fabric for current and future multi-/many-core implementations. However, the NoC has been shown to consume a significant fraction of the processor's power budget, up to 30% [11]. Hence, proper NoC power containment methodologies have to be considered.

Moreover, the march towards many-core microprocessors has been marred by the emergence of an ominous threat: waning *reliability* [19]. Transistors are more

D. Zoni (✉) · A. Canidio · W. Fornaciari
Politecnico di Milano, Milan, Italy
e-mail: davide.zoni@polimi.it; andrea.canidio@polimi.it; william.fornaciari@polimi.it

P. Englezakis · K. Chrysanthou · A. Prodromou · A. Panteli · C. Nicopoulos · Y. Sazeides
University of Cyprus, Nicosia, Cyprus
e-mail: englezakis.panayiotis@ucy.ac.cy; chrysanthou.kypros@ucy.ac.cy;
prodromou.andreas@ucy.ac.cy; panteli.andreas@ucy.ac.cy; nicopoulos@ucy.ac.cy;
yanos@cs.ucy.ac.cy

G. Dimitrakopoulos
Democritus University of Thrace, Xanthi, Greece
e-mail: dimitrak@ee.duth.gr

© Springer International Publishing AG, part of Springer Nature 2019   187
W. Fornaciari, D. Soudris (eds.), *Harnessing Performance Variability
in Embedded and High-performance Many/Multi-core Platforms*,
https://doi.org/10.1007/978-3-319-91962-1_9

susceptible to both permanent and transient faults. In addition to manufacturing and early life failures, future designs are anticipated to be increasingly vulnerable to aging and wear-out artifacts. Just like any on-chip component, the interconnection backbone is also affected by decreasing reliability [20]. A multitude of approaches to increase the fault tolerance and reliability of the NoC have been proposed. However, the vast majority of these approaches concentrate on fault *prevention* [7] and/or *recovery* [9, 14]. The equally important aspect of **fault detection and localization** has not been adequately addressed.

This chapter examines hardware-based monitors and knobs to tackle static power consumption, as well as to detect and localize faults occurring within the NoC. Specifically, two distinct solutions are proposed: the first—called *BlackOut* [27]—tackles static power consumption, and the second—called *NoCAlert* [6]—detects and localizes faults in the control logic of the router. Both solutions are autonomous, hardware-based mechanisms that operate at the micro-architectural level. The primary reason behind the decision to implement both techniques as *autonomous* monitor-and-knob mechanisms is the imperative need for swift action (operating at the granularity of only a few cycles), in order to minimize the impact on performance. Since the performance of the NoC is critical to the overall performance of the microprocessor, NoC routers operate with extremely low latencies (e.g., there are various state-of-the-art single-cycle NoC routers). Consequently, autonomous monitor-and-knob mechanisms that can operate at such ultralow latencies are highly desirable.

9.1.1   The BlackOut Architecture

The way conventional NoCs decide on the buffer depth is based on the "worst-case" scenario of all traffic going through a single virtual channel (VC). Specifically, the buffer depth of each *individual* virtual channel buffer is chosen as to cover the so-called credit round-trip time (RTT). Thus, each individual VC buffer is provisioned with this "worst-case" depth, which leads to overall over-provisioning from the perspective of the entire router. Since real traffic is spread among multiple virtual channels, most buffers are almost never filled up.

Furthermore, the amount of *different* virtual channel buffers that must be present in the NoC is typically dictated by the upper-layer cache-coherence protocol, which requires the different message classes (e.g., *request* and *reply* classes) to be separated, to ensure protocol-level deadlock freedom.

Since the NoC faces low traffic for the majority of the time, the overdesigned resources remain idle, or underutilized, thereby wasting energy due to leakage power consumption. The latter is swiftly emerging as a dominant power consumption component in current and future technology nodes.

Extensive prior research has attempted to tackle leakage power consumption in NoCs by exploiting the efficacy of power gating to dynamically switch off unused resources within the NoC [5, 15] or by carefully reducing its implemented hardware resources [17, 18, 28]. However, power gating introduces a non-negligible delay overhead to wake up the power-gated resources, thus posing a serious threat to overall system performance. Moreover, the traffic traversing the NoC cannot easily be predicted at run time, which makes predictive wake-up strategies ineffective. The design of effective power-gating methodologies for NoCs—aiming to aggressively reduce leakage power consumption—is usually achieved by means of complex and custom-designed mechanisms. Such architectures exploit additional gather networks to steer information to a centralized power-gating module [8], or they implement complex forwarding networks to timely provide prevailing traffic information to distributed power-gating modules [5].

BlackOut is a control-theory inspired power-gating methodology that dynamically switches off and on individual NoC buffers. Most importantly, the power savings are reaped with a negligible impact on overall system performance. The focus of *BlackOut* is on NoC architectures employing a fixed number of statically allocated VC buffer depths. A more detailed and in-depth description of this work can be found in [27].

9.1.2 The NoCAlert Architecture

Traditionally, fault detection is undertaken by built-in self-test (BIST) mechanisms. The BIST process may be executed by the manufacturer prior to shipment, or it may constitute part of system boot-up [12, 24]. Runtime BIST is also possible, but system operation is (partially) halted while the module-under-test is examined [9, 13].

Near-instantaneous fault detection may be achieved in the data path of the interconnect through the use of *error detecting codes*. While this methodology guarantees protection of the message contents, faults within the *control logic* of the NoC may still wreak havoc with the operation of the entire CMP. Hence, what is needed to guarantee functional correctness within the NoC is to *protect the NoC's control logic* (assuming that the flit *contents* are protected by error-detecting/error-correcting codes).

We hereby propose **a comprehensive online fault-detection and localization mechanism**, aptly called *NoCAlert*, which provides full fault coverage for all on-chip network control logic components and achieves *instantaneous* detection of any erroneous behavior caused by single-bit, single-event faults. More specifically, NoCAlert can detect (a) single stuck-at (permanent) faults, and (b) single-event transient faults. Moreover, the proposed mechanism is also able to pinpoint (localize) the fault with extremely high accuracy. The NoCAlert mechanism is based on the notion of *invariance checking*, whereby the system is continuously checked for illegal outputs as a result of upsets. An *illegal output* is defined here as an operational decision that violates the functional correctness rule(s) of a particular component. The underlying principle of this technique is inspired by prior efforts to protect the microprocessor by using invariances [16]. NoCAlert comprises several checker micromodules distributed throughout the NoC router. The checkers never interfere with—or interrupt—the operation of the NoC and they provide real-time online fault detection. In essence, NoCAlert implements extremely lightweight **real-time hardware assertions** that can detect illegal outputs within the NoC's control logic. Upon detection of a fault, NoCAlert can analyze the pattern of raised assertions to determine the fault location. A more detailed and in-depth description of this work can be found in [6].

9.2 BlackOut: Enabling Fine-Grained Power Gating of Buffers in Network-on-Chip Routers

Section 9.2 describes the *BlackOut* architecture in detail. Specifically, Sect. 9.2.1 details the assumed baseline router, which only represents a use case in this work. Note that *BlackOut* is applicable to any routing algorithm, NoC topology, and router pipeline depth. Section 9.2.2 concisely articulates the novel contributions of this technique. Section 9.2.3 describes related work in this field. The *BlackOut* architecture is described in Sect. 9.2.4, while Sect. 9.2.5 presents evaluation results and analysis.

9.2.1 The Assumed Baseline Network-on-Chip Router

This work focuses on a wormhole router that supports both virtual channels (VCs) and virtual networks (VNETs). A generic four-stage pipelined implementation is assumed with the following stages: (1) buffer write/route computation (BW/RC), (2) virtual channel allocation (VA), (3) switch allocation (SA), and (4) switch traversal (ST). A single additional cycle for link traversal (LT) is also assumed. As previously mentioned, the proposed BlackOut methodology is not constrained to a specific pipeline depth/micro-architecture; it can be used with any pipeline depth. The proposed solution accounts for the support of VNETs, which are required to enable coherence-protocol support at the NoC level. Traffic within one VNET is isolated from other VNETs, and packets traversing one VNET are not allowed to change VNET in-flight. A packet is split into multiple atomic transmission units, called *flits*. The first flit of each packet is the head flit, which opens up the wormhole. A body flit represents an intermediate flit of the original packet, while the tail flit is unique for each packet and it "closes" the packet and its wormhole. This work assumes the use of look-ahead RC (LRC) [10], i.e., the destined output port in the downstream router is pre-computed in the upstream router.

9.2.2 Novel Contributions

The *BlackOut* mechanism relies on efficient and fine-grained power gating of individual VC buffers within the router's input ports, in order to aggressively reduce leakage power consumption within the NoC. The VC buffers are dynamically switched off and on, based on prevailing traffic conditions, without impacting network (and thereby system) performance. The proposed architecture makes three major contributions to the state of the art:

- **Simple design**—*BlackOut* does not require complex and custom-designed forwarding networks to collect power-gating information.
- **Operational versatility**—*BlackOut* is not constrained to any particular NoC topology, nor to a specific routing algorithm, or a particular micro-architecture. It outperforms (in terms of energy savings) the current state-of-the-art methodologies under all traffic injection rates, by 35%, and it can reap energy savings of up to 70%, on a per-router basis.
- **Optimized micro-architecture**—*BlackOut* relies on a duet of key micro-architectural concepts (discussed in depth in Sect. 9.2.4) that operate synergistically and in unison: (a) *late binding*, and (b) *flow balancing*. The *late binding* technique allows the router to assign a particular buffer to a packet when the packet's head flit arrives at the downstream input port. The *flow balancing* concept is derived from the observation that traffic variations over short periods of time tend to be small. This implies that the buffer occupancy experiences slight changes over time, thereby allowing for a very effective fine-grained power-gating strategy. Overall, *BlackOut* operates on a *sliding window* concept, whereby the window includes all the active (switched on) buffers, and the window may wax and wane by switching on and off additional buffers, respectively.

9.2.3 Related State-of-the-Art Power-Gating Solutions

The *BlackOut* technique is compared against two state-of-the-art power-gating solutions. We characterize these solutions based on their operational granularity, either at the router-level, or at the buffer-level.

9.2.3.1 Power Gating at a Router-Level Granularity

The Power Punch [5] methodology promises non-blocking power gating in NoC routers. It tries to completely hide the router's wake-up latency of eight cycles by sending wake-up signals up to three hops in advance, adding the capability of hiding a variable number of cycles depending on the router's pipeline. However, it is quite costly to have a wire from every router to every router three hops away dedicated to signal wake-ups. Therefore, Power Punch proposes a way to encode wake-up signals to minimize this complexity. The encoding wires are additional to the router-to-router link wires. Power Punch is **the current state of the art in terms of router-level power-gating granularity**. As a result, it is one of the two techniques that will be compared against *BlackOut* later on in Sect. 9.2.5.

9.2.3.2   Fine-Grained Power Gating

Ultrafine-grained runtime power gating [15] discusses a power-gating methodology considering an NoC with only a single VC per VNET. Different parts of the same router can be dynamically power-gated to reduce leakage power. It organizes the router in micro-power domains that can be independently power-gated: buffers, VA, crossbar, and output latches. A critical contribution of [15] is the complete RTL design at 65 nm. In particular, the wake-up time for a 128-bit 4-slot buffer is estimated to be 2.8 ns. This result is aligned with [29], where a 4-slot 32-bit buffer is woken up in 1 ns. The wake-up policy relies on look-ahead signals between routers to hide the VC wake-up latency. This work is the most suitable candidate for a comparison against *BlackOut*. It is **the state of the art in buffer-level power gating**, and it is directly comparable to *BlackOut*. Hence, a detailed comparison between the two techniques will be provided in Sect. 9.2.5. The proposed *BlackOut* mechanism relies on a novel *flow balancing* concept, which can dynamically predict the upcoming demand for buffer space. Thus, *BlackOut* can proactively switch on buffers, thereby minimizing the impact on performance. Additionally, the new mechanism uses a dynamic virtual channel buffer remapping technique, which minimizes the number of switch-on and switch-off events.

9.2.4   The BlackOut Methodology

This section describes the details and key attributes of the proposed *BlackOut* mechanism, which is a control-theory inspired methodology that power-gates individual input VC buffers in NoC routers to reduce the static power consumption. Said methodology is able to aggressively power off the majority of the input buffers in a router, while incurring negligible overhead to the overall system performance.

BlackOut is implemented on top of credit-based flow control, with two main design constraints. First, the changes are minimal and the design is very lightweight. Second, *BlackOut* directly exploits the flow control mechanism, so that it is guaranteed not to steer the system into a deadlock, as long as the baseline system is guaranteed to be deadlock-free.

The *BlackOut* design methodology is split into two parts and encompasses a generic pair of actors: the upstream and the downstream modules. The *physical actuation* is performed by the downstream module, i.e., the router input port where the buffers are situated. Conversely, the upstream module is in charge of the *control-*

and-command dispatches, namely signaling to the downstream module to switch off or on a particular buffer. This action is performed by the router's output port, or by the network interface controller (NIC). Therefore, each router implements one instance of the upstream module policy per output port, and one instance of the downstream module policy per input port. Instead, the NIC only requires an implementation of the upstream module policy in its output port, since the NIC is never considered to be a downstream module.

BlackOut manages the on/off state of buffers by employing a **sliding window**. At any point in time, a subset of the available buffers in the input port are idle, and they can potentially be switched off. Depending on the traffic conditions, the sliding window is incremented, or decremented, by switching on or off a new buffer, respectively. Thus, a window can be fully closed (when all buffers are switched off), or completely open (when all buffers are switched on). *BlackOut* is founded on two different operational pillars, which provide flexibility and great power-saving capabilities with negligible performance overhead: (a) *flow balancing* and (b) *late binding*.

Without loss of generality, the rest of this section considers an upstream router output port (OP) and a downstream router input port (IP) pair. The *flow balancing* concept applies to the upstream router output port and is linked to the inherent balance between the current buffer occupancy in the downstream input port and the one in the near future. In particular, the SA requests in the upstream router—that are routed to the OP—represent the buffers currently in use in the downstream router IP. They passed the VA stage, so a buffer is allocated in the downstream router for each of them. Conversely, the portion of requests in the upstream router that are routed to the OP and are in the VA stage represent the future buffer requirements. The VA requests directed to the OP can be more than the SA requests, thereby implying that the number of active buffers cannot cope with the packet flow. On the other hand, if the VA requests are lower than the SA requests for the considered OP, then the number of active buffers can manage the VA requests in the future without the need to switch on other buffers. In other words, the *flow balancing* concept defines the OP state as the current state, i.e., the active buffers or SA requests, and the current VA requests as the future state. This basic information is used to develop a policy to increase, or decrease, the number of active (on) buffers in the corresponding downstream IP.

The *late binding* technique optimizes the physical allocation of a packet to a buffer in the downstream IP. In an NoC, the virtual and physical allocation of a buffer to a packet happens at different times, since the VA in the upstream router takes place several cycles before the BW in the downstream one. *Late binding* is a mechanism to decouple virtual allocation of the channel from physical binding, eventually remapping a virtual channel onto a different physical buffer. This means that the upstream VA exploits the baseline flow control policy, while the head flit of the packet is linked to a particular physical buffer only when it actually arrives at the IP of the downstream router. A buffer is then assigned, depending on the current availability and state (on/off) of the buffers in the IP. It is worth noting that a packet arrives at the downstream IP if the VA in the upstream node is successful. Hence, no deadlock or lack of resources can result from *late binding*, since no speculation is involved. Only a buffer mapping occurs at the moment of arrival.

The depth of the router's pipeline can mask a number of the cycles needed to wake up buffers in the downstream router. A four-stage pipeline (as the one used in our simulations) can mask up to two cycles of wakeup latency. If the wakeup latency of a buffer is more than that, then the upstream router would need to stall its pipeline. As the pipeline depth shortens, this phenomenon intensifies. For example, a two-stage router can mask no wakeup latency.

9.2.5 Evaluation and Results

The assessment of the *BlackOut* methodology of Sect. 9.2.4 considers performance and energy savings. This evaluation is distributed over the following four subsections. Section 9.2.5.1 describes the simulation setup. Performance and energy simulations considering both synthetic traffic and real applications are then presented in Sects. 9.2.5.2 and 9.2.5.3, respectively.

9.2.5.1 Experimental Setup

BlackOut has been fully implemented and integrated into an enhanced version [30, 31] of the *gem5* cycle-accurate simulator [2], extended in [30, 31] to also include a cycle-accurate model of power gating in the NoC. DSENT [25] is used to extract power data of the simulated NoC architectures. *BlackOut* is compared against two state-of-the-art solutions at two different granularities. The coarse-grained solution is *Power Punch* [5], while the fine-grained technique is UFG [15]. Both mechanisms were discussed extensively in Sect. 9.2.3.

Two different synthetic traffic patterns are considered, i.e., uniform random and tornado, as well as nine SPLASH-2 [26] benchmarks configured to allow a maximum of 64 parallel threads. Detailed parameters of the simulations are reported in Table 9.1.

The latency to wake up a single buffer (T_{on}) is a critical parameter that can outweigh any benefit of the methodology. *BlackOut* is evaluated considering three different latencies to wake up a single buffer (T_{on}), i.e., 1, 2, and 4 cycles. The same parameters are also used when evaluating the state-of-the-art fine-grained power-gating methodology presented in [15]. For *Power Punch* [5], the T_{on} latency is constant at eight cycles.

9.2.5.2 Results Using Synthetic Traffic

This section discusses the latency, saturation point, and energy profile of the *Black-Out* proposal, considering two different synthetic traffic patterns: uniform random

Table 9.1 Experimental setup: salient parameters pertaining to the processor, router micro-architecture, assumed technology, and employed traffic patterns

Processor core	1 GHz, Out-of-order Alpha core
L1I cache	2-way, 32 kB
L1D cache	2-way, 32 kB
L2 cache	8-way, 256 kB per bank
Coherence prot.	MOESI (3-VNET protocol)
Router	4-stage wormhole
	32-bit link width
	Buffer depth four flits
	Two virtual channels (VCs) per VNET
	Three VNETS (Garnet network [1])
Topology	2D-mesh/8×8 (1 core/tile)
Routing Alg.	Deterministic XY
Technology	45 nm at 1.0 V
Synthetic traffic patterns	Uniform random, tornado
Real traffic	SPLASH-2 benchmarks [26]

and tornado. The *BlackOut* framework is compared against (a) the current state-of-the-art buffer-level (i.e., fine grain) power-gating methodology presented in [15] (henceforth identified as the *Ultrafine-Grain, UFG,* power-gating scheme); (b) the current state-of-the-art router-level (i.e., coarse grain) power-gating methodology of *Power Punch* [5]. The energy savings reported are normalized to the baseline NoC that has no switching-off capabilities, and, therefore, no performance penalties are introduced.

The aim of this subsection is two-fold: (1) to demonstrate that router-level (coarse-grain) power-gating proposals lack flexibility, since even in the presence of a single traffic flow from one input–output port pair, the router has to be kept switched on; (2) to demonstrate that *BlackOut* can provide fine-grained control over the switched off resources, thereby extracting better performance under all injection rates up to the network's saturation point.

Realistic traffic comprises two different packet sizes: (1) *single-flit* packets used for control/coherence messages and cache/memory *requests*, and (2) *multi-flit* packets used for cache/memory *reply* messages (i.e., the packets that contain the actual cache line requested). In our evaluation, the size of the multi-flit packets was explored further, providing results with 1-flit, 3-flit, 5-flit, and 9-flit *data* packet sizes (in separate simulations).

Each simulated scenario considers three distinct VNETs to better resemble the setup required by a realistic cache-coherence protocol in multi-core environments. Two VNETs are always considered to inject single-flit packets, while the third one can inject multi-flit packets (as described above), depending on the selected configuration.

All the figures in the rest of this section share the same format, showing the injection rate in flits/node/cycle on the *x-axis*. The total latency (i.e., network latency

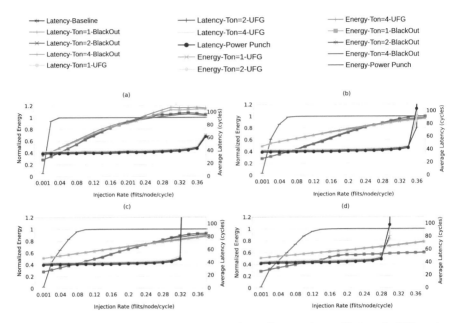

Fig. 9.1 Latency/throughput and energy results using **uniform** random traffic in an 8×8 2D mesh, with four different data-packet sizes (1, 3, 5, and 9 flits). Three architectures are compared: *BlackOut*, *Ultrafine-Grain (UFG)* [15], and *Power Punch* [5]. (**a**) Uniform random, data-packet size = 1 flit. (**b**) Uniform random, data-packet size = 3 flits. (**c**) Uniform random, data-packet size = 5 flits. (**d**) Uniform random, data-packet size = 9 flits

plus queuing latency) and the normalized energy consumption are depicted on the left and right *y-axes*, respectively.

Figure 9.1 reports the obtained results under uniform random traffic. All three techniques investigated (*BlackOut*, *UFG*, and *Power Punch*) show a saturation point almost identical to the baseline NoC, regardless of the data-packet size, thus highlighting the negligible impact of the power-gating actuator. However, *Power Punch*'s energy profile saturates at 100% (i.e., no energy savings) from very low traffic injection rates, whereas *BlackOut* has an energy profile that grows linearly with the injection rate. This implies that *Power Punch* can obtain the same performance as the baseline NoC, but without any energy savings, except when the injection rate is extremely low, i.e., below 0.04 flits/cycle/node. On the other hand, *BlackOut* provides better performance per consumed energy, since it can accurately maintain the minimum amount of active (switched on) buffers, without incurring any noticeable performance overhead. Note that even at the network's saturation point, *BlackOut* is still slightly below 100% of the normalized energy consumption for multi-flit data packet sizes of 3, 5, and 9. This is due to the fact that the depicted results are averaged over the 64 routers, and even if the input port of the traffic generator saturates, the internal NoC router ports are still able to keep some buffers switched off (albeit for a short period of time).

When all traffic consists of single-flit packets, both *BlackOut* and *UFG* consume more energy than the baseline NoC. This is due to elevated numbers of switch-on events. Since switch-on events consume extra energy, an increased number of such events lead to overall energy consumption that is higher than the baseline NoC. When the size of the multi-flit data packets increases to three and five flits, *BlackOut* outperforms *UFG* by up to 8%. Close to saturation, the two techniques behave similarly, yielding approximately the same amount of energy savings (with *UFG* performing marginally better at saturation). However, the NoC is typically never operated at (and beyond) its saturation point, since performance deteriorates to unusable levels. When the size of the multi-flit data packets increases to nine flits, *BlackOut* yields 15% more energy savings, on average, than *UFG*.

In terms of performance, *BlackOut* and *UFG* behave similarly, incurring a negligible increase in network latency. Overall, *BlackOut* yields 29% higher energy savings, on average, than *Power Punch*, and 8% than *UFG*, while incurring a near-negligible performance overhead of 2%.

Note that *Power Punch* can save more energy than *BlackOut* (due to the former's ability to completely switch off a router, instead of individual buffers) only at extremely low injection rates, i.e., below 0.01.

The results obtained using tornado synthetic traffic are reported in Fig. 9.2. All three mechanisms—*BlackOut*, *UFG*, and *Power Punch*—achieve the same

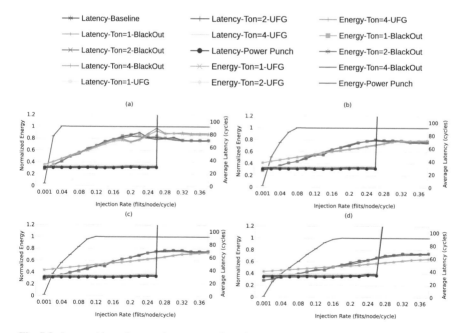

Fig. 9.2 Latency/throughput and energy results using **tornado** traffic in an 8×8 2D mesh, with four different data-packet sizes (1, 3, 5, and 9 flits). Three architectures are compared: *BlackOut*, *UFG* [15], and *Power Punch* [5]. (**a**) Tornado, Data-packet size = 1 flit. (**b**) Tornado, Data-packet size = 3 flits. (**c**) Tornado, Data-packet size = 5 flits. (**d**) Tornado, Data-packet size = 9 flits

saturation point as the baseline NoC. Once again, for injection rates higher than 0.06 flits/node/cycle, *Power Punch* provides no energy-saving benefits. On the other hand, *BlackOut* is flexible enough to exploit the traffic characteristics and reap energy savings of around 40%, even at the saturation point. In other words, at the saturation point, *BlackOut* consumes only 60% of the energy consumed by the baseline NoC. This substantial energy saving is attributed to the tornado traffic pattern, which mainly stresses horizontal (East/West) inter-router links in a 2D mesh, thereby allowing *BlackOut* to switch off the buffers in the vertical (North/South) input/output ports. Similar to the uniform random traffic results, *BlackOut* again shows better flexibility than *Power Punch*, irrespective of the data packet size. As a result, *BlackOut* achieves almost the same performance as the baseline NoC, but with massive energy savings (under all injection rates up to the saturation point). Note that *UFG* can sometimes reap even higher—albeit not by much—energy savings than *BlackOut* under tornado traffic. In summary, under tornado traffic, *BlackOut* yields 28% higher energy savings, on average, than *Power Punch*, while containing the performance overhead to merely 3%.

9.2.5.3   Results Using SPLASH-2 Benchmark Applications

This section evaluates the performance and energy-saving attributes of *BlackOut* when using real application workloads from the SPLASH-2 benchmark suite [26]. We still consider an 8×8 2D mesh multi-core setup. Similarly to the previous subsection, the *BlackOut* framework is compared against all state-of-the-art power-gating methodologies, i.e., *UFG* [15] and *Power Punch* [5].

The goal of this section is to demonstrate that *BlackOut* can outperform state-of-the-art buffer-level power-gating techniques, due to its sophisticated, yet simple, micro-architectural design features, as described in Sect. 9.2.4. Performance results are not provided here, since—as it will be mentioned later on—the performance impact is minimal.

We compare the three competing mechanisms using real applications. Regarding these experiments, the T_{on} latencies used are the same as before. The simulations revealed that the performance overhead observed with the *BlackOut* and *UFG* techniques is limited to within 3%. Also, as the T_{on} latency is increased from 1 to 4 cycles, only a negligible increase in the performance overhead is observed. *Power Punch* behaves as expected, i.e., it does not suffer from any performance degradation, as described in [5].

Figure 9.3 presents the energy consumption results, normalized to the baseline NoC. *BlackOut* achieves, on average, 36% lower energy consumption than *UFG*. On the other hand, *Power Punch* exhibits very high variability in its energy-saving effectiveness, based on the application's traffic behavior (see Fig. 9.3). More importantly, *BlackOut's* behavior is very consistent (with minimal variance) throughout all benchmarks (energy savings ranging from 70% to 73%), while *Power Punch* yields widely varying energy savings, ranging from 39% all the way to 94%. This variability is based on the variability of the injection rates observed in the

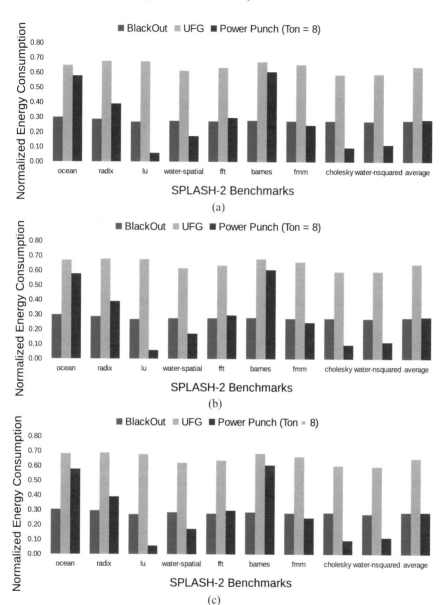

Fig. 9.3 **Energy consumption** results of *BlackOut*, *UFG*, and *Power Punch* when using SPLASH-2 benchmark applications [26]. The results are normalized to the baseline NoC. Three different T_{on} latencies (i.e., the number of cycles required to switch on a buffer) are considered for the *BlackOut* and *UFG* methodologies: 1, 2, and 4 cycles. *Power Punch* is only simulated with a T_{on} latency of eight cycles. (a) $T_{on} = 1$ cycle (for *BlackOut* and *UFG*). (b) $T_{on} = 2$ cycles (for *BlackOut* and *UFG*). (c) $T_{on} = 4$ cycles (for *BlackOut* and *UFG*)

SPLASH-2 benchmark suite. Applications on the lower end of the injection rate spectrum favor *Power Punch*. However, *BlackOut* is generally more consistent.

9.3 An Online and Real-time Fault-Detection and Localization Mechanism for Network-on-Chip Architectures

Section 9.3 describes the NoCAlert architecture in detail. Specifically, Sect. 9.3.1 describes the concept of *invariance checking* within the NoC. Section 9.3.2 describes the NoCAlert framework. Section 9.3.3 describes the employed experimental/evaluation methodology, while Sect. 9.3.4 presents and analyzes the evaluation results.

9.3.1 Invariance Checking Within the Network-on-Chip

The NoCAlert mechanism is based on the concept of *invariance checking*. The system is continuously examined for illegal outputs as a result of some kind of fault. The term *illegal output* is defined here as an operational output that is impossible to occur, based on the set of functional correctness rules of a given component. The assertions are implemented in *hardware* to provide fast detection of anomalies. The pattern of raised assertions is then used to pinpoint the location of the detected fault.

The salient characteristic of invariance checking is the fact that only *functionally illegal* outputs are flagged as violations. A fault that causes the generation of an erroneous, yet functionally legal, output will *not* be identified as a breach of correctness. What we aim to explore in this work—among others—is how often, and under what conditions, such non-invariant faults could potentially lead to compromised network-level correctness.

The router architecture used is the same as the one used in Sect. 9.2.1, with the only differences being that atomic buffers are used here, and no look-ahead routing is employed. In this work, invariances were constructed by observing the operation and behavior of each functional module. The completeness of invariances depends on the completeness of the functional analysis of the design itself. If the invariance checkers cover all functional rules, NoCAlert will detect *any illegal* behavior. In the case of the baseline NoC router micro-architecture, a total of 32 invariance types (categories) were identified and documented in [23] to detect transient faults, with 3 more added here to detect permanent faults.

In total, the 35 invariance categories translate to 35 extremely lightweight checker modules. The checker modules' outputs are not registered (i.e., stored); they are fed directly to the localization logic of NoCAlert, as will be explained later on.

9.3.2 NoCAlert: An Online, Hardware Assertion-Based Fault-Detection and Localization Mechanism for Networks-on-Chip

The proposed NoCAlert framework utilizes the principle of invariance checking and implements it in the form of *real-time hardware-based assertions*. The key idea is to have a simple *hardware* checker module for every NoC component. This checker module will take as inputs the inputs and outputs of the protected component and it will check whether any functional rule is broken during the component's operation.

Two single-bit, single-event fault models are employed in this work: (a) transient and (b) permanent stuck-at faults (stuck-at-0 and stuck-at-1).

9.3.2.1 Ensuring Network Correctness Using Invariances

Prior research [3, 22] has identified four main conditions that *ensure* functional correctness within the network: (1) no packets are dropped, (2) delivery time is bounded, (3) no data corruption occurs, and (4) no new packet is generated within the network. These four conditions guarantee functional correctness [3, 22]. The 32 identified invariances were categorized according to the aforementioned four general requirements, as described in [23]. The three new invariances target the second requirement, i.e., *bounded delivery*, since they mainly monitor for packets/flits that may get stuck in some router due to a fault.

9.3.2.2 Hardware Complexity of the NoCAlert Checkers

The NoCAlert fault-detection mechanism consists of an array of distributed hardware checkers, which constantly monitor the modules comprising the control logic of the router. Each checker is a simple combinational circuit performing a specific check, according to the rules of the module being monitored. The NoCAlert framework is not bound to any specific router micro-architecture. Any alterations to key router components, or the addition of new modules within the router will simply generate new invariances, based on the modified/new functional specifications.

9.3.2.3 Faults That Do Not Cause Invariance Violations

As previously mentioned, invariance checking only detects *illegal* outputs, not necessarily *incorrect* ones. Faults that give rise to *functionally legal* outputs will not be detected. The two elemental questions here are the following:

- If such non-invariant upsets cause some other functional/invariance violation *later on* in the network, will the fault be caught by one/some subsequent NoCAlert checkers?
- If these non-invariant upsets do *not* cause any other functional/invariance violation *later on* in the network (i.e., they are never caught by any NoCAlert checker), do they end up affecting the overall network correctness (as defined in Sect. 9.3.2.1 and [3, 22])?

The extensive simulations of Sect. 9.3.3 will answer these two important questions. It turns out (empirically) that all the non-invariant faults that end up causing a functional error later on are, indeed, successfully captured by subsequent NoCAlert checkers, whereas the non-invariant faults that do not cause any other invariance violation later on turn out to be benign, as far as overall network correctness is concerned.

9.3.2.4   Enabling Fault Localization

To minimize the impact on performance and the disruption in the system's operation, two NoCAlert delay elements must be minimized: (a) the delay between the manifestation of a fault and its detection, and (b) the delay between detection and localization. Most importantly, ambiguous (imprecise) fault localization leads to overly pessimistic/conservative solutions. For example, the inability to pinpoint a single permanently faulty component within an NoC router may lead to the disconnection of the entire router from the network. Given a quick and accurate fault-localization unit, recovery mechanisms can be fine-tuned accordingly to yield a near-optimal—with respect to performance degradation and service disruption—fault-tolerant system.

To enable fault localization, NoCAlert utilizes the output of its detection checkers. The outputs of all checker modules within a router may be **grouped into a single *vector***, whereby each bit corresponds to the output of a single invariance checker. Whenever a checker detects a fault, this bit vector is updated to reflect the disturbance. Essentially, the fault-detection vector can be viewed as a *fault detection trace*, which, according to the generated vector pattern, could potentially yield the location of the fault.

The ***Localization Unit*** introduced here—one per router—tries to tackle the issue of identifying (i.e., localizing) the fault-afflicted NoC router component, by analyzing the assertions of the NoCAlert checkers that have detected the fault. The output bits of all NoCAlert checker modules within each router are fed directly into the router's Localization Unit and are not latched at the end of each cycle. As soon as an assertion is raised by any one of the checker modules within a router, the Localization Unit of said router is triggered into action; that is, it generates a so-called ***Assertion Vector***, which includes the status (asserted/de-asserted) of all checkers in the local router at the instant of the first raised assertion (i.e., the instant of first detection). The size of the Assertion Vector for the 5-port baseline router is 300 bits. This Assertion Vector is then used by the Localization Unit to identify the fault's location within the router.

Each Localization Unit operates using only the assertions of a single router. Thus, the localization process is distributed across routers, and each one of the localization units is *independent*, i.e., does *not* need to communicate with other routers' units to identify the fault location. This valuable attribute is a direct consequence of the instantaneous fault-detection capability of NoCAlert, which implies that the detection assertions are spatially close to the fault location. Experimental evaluations later on will validate this NoCAlert property. In general, fault localization is the process of arbitrarily dividing a module, e.g., the NoC router, into multiple sectors and then pinpointing the fault location to within a particular sector. The size and granularity of these sectors will be discussed later on. The Localization Unit consists of a combinational circuit, which calculates whether one of the aforementioned sectors could be the faulty one. This circuit is empirically created, based on a large number of fault-injection simulations. For each simulation, a pair is created, which correlates the specific Assertion Vector created at the instant of detection (henceforth termed the *first-detection cycle*) with the location where the fault was injected. Each distinct assertion vector is given a unique Assertion Vector ID for identification purposes; that is, the ID is used to refer to a specific Assertion Vector, which is matched to a specific fault location. All these pairs are then analyzed to deduce signatures that could identify whether a fault happened at a specific location. The process of building and optimizing these pairs/correlations is heuristic-based and makes use of the *Espresso* logic minimizer tool [4]. Initially, all fault-injection simulations are inspected. For every simulation that results in a fault detection (i.e., at least one checker module is asserted), a pair is generated. The pair consists of (1) the location of the fault (as injected in that specific simulation), and (2) the corresponding Assertion Vector created in the first-detection cycle of that simulation.

Based on these generated pairs, when the Localization Unit receives one of these Assertion Vectors as input, it will output the respective fault location. The size (in bits) of the Localization Unit's output depends on the localization *granularity*, as will be explained later on. This output is registered, so that it can be subsequently used for diagnostic and/or recovery purposes by another mechanism (if it exists), or a higher level in the system stack.

The complexity of the localization unit is a trade-off between localization *ambiguity* (imprecision) and *granularity*. Ambiguous localization occurs when the localization unit yields more than one possible fault locations, i.e., the fault may have occurred in any one of two (or more) locations. The reason for such ambiguity in the localization unit's output is the nonuniqueness of the Assertion Vector itself. Localization granularity refers to the size of the reported fault location. This is related to the size of the "sectors" described previously. NoCAlert framework supports three different localization granularities. The coarsest granularity (*router-level*) localizes faults that occurred in a specific router. The second granularity (*pipeline-stage-level*) tries to identify the pipeline stage that experienced a fault. The finest granularity (*port-level*) tries to localize the faults down to a single input port of the router.

The size (in bits) of the output of each Localization Unit depends on the chosen localization granularity, and it varies from 5 (coarsest) to 44 (finest). In the absence of localization ambiguity, the output is a one-hot bit vector, which uniquely identifies the fault location (at the supported granularity level). On the other hand, an ambiguous output is a bit vector that is *not* one-hot, with each "1" indicating a unique (and separate) possible fault location.

The Localization Unit relies on the analysis of the Assertion Vector, which is generated from the outputs of the hardware checkers (i.e., the detectors). However, not all bits in the Assertion Vector are always necessary to yield a fault location. This fact allows us to perform several optimizations, in order to minimize the information processed by the Localization Unit. This, in turn, simplifies the entire localization logic. Specifically, we employ two optimizations: (a) compacting the Assertion Vector, and (b) discarding portions (i.e., pruning) of the Assertion Vector for each localization granularity level.

When localization happens at the router- or pipeline-stage-level granularities, the specific instance of the asserted hardware checker is not significant. In many cases, the only required information is the specific *invariance type* (category) to which the asserted checker output bit belongs. Thus, the Assertion Vector can be reduced to include only one "summary" bit for some invariance types.

Moreover, since the Localization Unit is created based on a set of empirically generated pairs of Assertion Vectors and corresponding fault locations, portions of the Assertion Vector might, in fact, be redundant. For example, several checkers have the same value for all faults injected in a particular location (i.e., some checkers are either always asserted or always de-asserted). The employed pruning process carefully analyzed the contribution of every single entry in the Assertion Vector to identify the "utility" of each checker with respect to the localization process. Pruning also took care of situations where two (or more) checkers were always simultaneously asserted. Thus, using only one of those checkers would be enough for localization. Eventually, the pruning process yielded a smaller Assertion Vector that only included the checkers that were responsible for correct and accurate fault localization. Specifically, the original 300 bits of the Assertion Vector were pruned to 170 bits (for the *router*-level localization granularity), or 225 bits (for the *pipeline-stage*-level and *input/output-port*-level localization granularities).

9.3.3   Experimental Methodology

9.3.3.1   Experimental Setup

The goal of the experimental evaluation is to thoroughly assess the *efficacy* and *efficiency* of the NoCAlert mechanism in a realistic environment. Our evaluation approach is double-faceted, covering two abstraction levels: (a) C++ NoC simulations at the network level, and (b) simulations at the **RTL netlist-level**.

For the *network-level* evaluation, the cycle-accurate GARNET NoC simulator [1] is employed. Synthetic traffic patterns were used to thoroughly test the 8×8 mesh and accurately capture the salient characteristics of the design. The NoCAlert framework is also compared to ForEVeR [21, 22]. The ForEVeR framework [21, 22] complements the use of formal methods and runtime verification to ensure functional correctness in NoCs. While ForEVeR's goal is to protect against escaped design-time verification errors with a runtime technique, the scheme may also be used to provide robustness against runtime faults. ForEVeR relies on time *epochs* and counters, which—if needed—trigger a recovery mechanism, which delivers the in-flight data to the intended destination via an additional lightweight checker network that is assumed to be 100% reliable.

The router is five-stage pipelined (4 intra-router stages + 1 link traversal stage), with four 5-flit deep VCs per input port, and 128-bit inter-router links. Atomic VC buffers, wormhole switching, and credit-based flow control are also assumed. The routing algorithm used is deterministic XY.

For the hardware evaluation and RTL simulation parts, we implemented the baseline NoC router augmented with the NoCAlert detection and localization mechanisms in Verilog Hardware Description Language (HDL). The resulting design was synthesized using Synopsys Design Compiler and 65-nm commercial TSMC libraries at 1-V operating voltage and 1-GHz clock frequency.

9.3.3.2 Evaluation Framework

For the network-level evaluations, faults were injected within the NoC routers at the **fine granularity of individual sub-components** (VA and SA arbiters, RC unit, etc.). Our model has the capability of *injecting single-bit faults at the inputs and the outputs of each individual module*. One simulation run at a single traffic injection rate and one network state consists of 35,424 different simulations (to exhaustively inject faults in all 11,808 possible locations of an 8×8 mesh, assuming the specific fault models used in this work, i.e., transient bit-flips, and permanent stuck-at-0 and stuck-at-1 faults). The traffic injection rate was varied from low to high (0.1–0.4 flits/node/cycle) in steps of 0.05 flits/node/cycle. Moreover, three different scenarios of fault-injection instances were studied (fault injection at cycle 0, 32,000, and 64,000). Hence, 21 different scenarios were investigated (7 injection rates $\times$ 3 injection times), for a total of $21 \times 35,424 \approx 744,000$ fault-injection simulations.

The exact same experiments were also run in a *fault-free* environment and detailed flit ejection logs were collected and compiled in a so-called *Golden Reference* (GR) report. The GR is then used to ensure that no violations of the four network correctness rules of Sect. 9.3.2.1 and [3, 22] occur. Furthermore, the GR also detects any changes in the intra-packet flit order (such order violations constitute erroneous behavior). Since NoCAlert only captures faults that cause invariance violations, the GR is used to facilitate the investigation of the two key questions posed in Sect. 9.3.2.3. Moreover, by comparing the GR with the equivalent *under-fault* log report, we can study the effects of any fault occurrence on overall network correctness. This allows us to assess the *false positive* (assertions that prove benign) and *false negative* (undetected network correctness violations) performances of both NoCAlert and ForEVeR [21, 22].

For the RTL netlist-level evaluations, faults were injected within the logic of each sub-module, and not just at their inputs and outputs. Note that the purpose of the RTL netlist-level analysis is the following: *any* internal fault (i.e., a fault injected within a module, not only at its inputs/outputs) should manifest as one of two cases: (1) masked fault (by internal logic), which would yield no error; or (2) error at a module's output, which would be captured by the checker (i.e., the case covered by the input/output approach of the network-level evaluation). The number of fault-injection locations at the RTL netlist-level was about 22,000, as compared to the 205 locations available in our GARNET-based (network-level) simulator.

9.3.4   Evaluation Results

9.3.4.1   Network-Level Simulation Results

As discussed in Sect. 9.3.3.1, simulation experiments were performed in an 8×8 2D mesh network using synthetic traffic patterns. In this subsection, we present a quantitative analysis of NoCAlert's efficacy and efficiency, as well as a comparison with the ForEVeR [21, 22] framework.

It is important at this point to differentiate the *injected faults* from the *actual errors* manifesting themselves at the network-level (as defined in Sect. 9.3.2.1 and [3, 22]). NoCAlert's ultimate goal is to ensure that no *actual error at the network-level* escapes detection. Therefore, injected faults that do NOT cause a real functional error within the network are viewed as benign. Based on this crucial differentiation, we classify each of NoCAlert's detection outcomes into one of the four main categories:

- **True Positive**: Event detected by NoCAlert when the injected fault causes an actual error at the network level (network correctness violation).
- **False Positive**: Event detected by NoCAlert when the injected fault turns out to be benign.
- **True Negative**: Nothing detected by NoCAlert when the injected fault turns out to be benign.
- **False Negative**: Nothing detected by NoCAlert when the injected fault causes an actual error at the network level (network correctness violation).

The simulation results under both examined fault models are summarized in Figs. 9.4 and 9.5. Specifically, Fig. 9.4 presents a breakdown of the fault-detection performance at two different fault-injection instances: cycle 0 (empty NoC), and cycle 32K (warmed-up NoC). The results at cycles 32 K and 64 K are very similar, so the 64 K results are omitted for brevity. Figure 9.5 shows the cumulative fault-detection delay distribution for True Positive faults. Overall, the evaluation of the

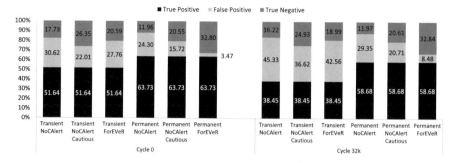

Fig. 9.4 Fault coverage breakdown (over all injected faults) using synthetic (uniform random) traffic in an 8×8 mesh at two different fault-injection instances (cycle 0 and cycle 32 K). The "NoCAlert Cautious" bars refer to a system where low-risk invariances are not immediately flagged as errors. Instead, the error is flagged only if further evidence indicates a problem [23]

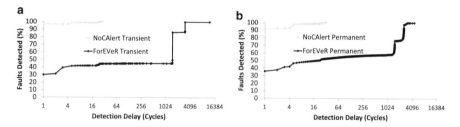

Fig. 9.5 Cumulative fault-detection delay distribution for True Positive faults under (**a**) transient and (**b**) permanent faults. The epoch duration in ForEVeR was set to 1500 cycles

detection capabilities of the two architectures under comparison yields five key observations:

1. NoCAlert registered *zero* false negatives, under both fault models as did ForEVeR [21, 22].
2. Transient and permanent faults exhibit the same trends, with the difference being that permanent stuck-at faults give rise to a higher percentage of True Positives, since the persistence of permanent stuck-at faults is much more likely to lead to an error.
3. As can be seen in Fig. 9.4, the true positive percentages are identical for both NoCAlert and ForEVeR, since they both detected *all* network correctness violations. There is a notable difference in the *False Positives*. ForEVeR's epoch-based approach allows for some benign faults to be masked.
4. NoCAlert provides near-instantaneous fault detection, with 97% of all true positive transient faults and 90% of all true positive permanent faults captured at the instant of manifestation (same cycle), as depicted in Fig. 9.5, *several orders of magnitude* lower fault-detection latency than ForEVeR [21, 22].
5. Injected faults that do *not* cause any invariance violation in the network are *always* benign (i.e., they never cause any network correctness violation).

Moreover, non-invariant upsets that cause a subsequent invariance violation are always successfully captured by NoCAlert. These fundamental results answer the two key questions posed in Sect. 9.3.2.3.

9.3.4.2 Fault-Localization Evaluation

NoCAlert's Localization Units only utilize the Assertion Vector generated at the instant of detection. Further, by looking only at the first-detection cycle, localization can exploit the fast detection times and ascertain that the fault has not had the time to cause widespread system malfunction. As illustrated in Fig. 9.6, which combines the results of both transient and permanent fault models, over 99% of the assertions raised during the first-detection cycle are within the affected router, and *all* assertions are at most one hop away. The following analysis is based on all the simulation results from both examined fault models (permanent and transient).

Figure 9.7 summarizes the localization capabilities of each NoCAlert hardware checker module. The x-axis lists the 35 NoCAlert invariance types, while the y-axis indicates the fault locations that each checker can help identify, based on all conducted experiments. It is evident in Fig. 9.7 that some checkers can, by themselves, determine the fault location, with no need to inspect other checker outputs.

NoCAlert's Assertion Vector allows for a fast and highly accurate localization solution. As discussed in Sect. 9.3.2.4, localization is achieved by identifying "signatures" (patterns) inside the Assertion Vectors that uniquely correspond to a specific fault location. When operating at the coarsest localization granularity (router-level), the output of the Localization Unit is unambiguous and 100% accurate. The output is also unambiguous and 100% accurate when operating at the pipeline-stage-level localization granularity. Ambiguity (aliasing) in the Localization Unit's output appears only at the finest localization granularity, the port-level. Figure 9.8 presents the probability of ambiguity in the output of the

Fig. 9.6 Distance distribution of raised assertions from the faulty router at the moment of detection

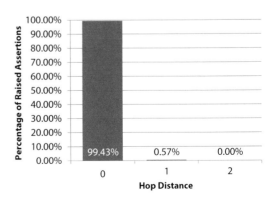

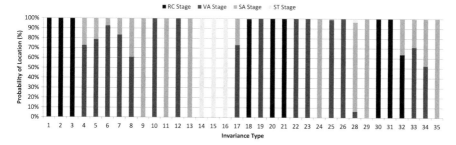

Fig. 9.7 Localization capabilities of each NoCAlert invariance checker module. The fault locations that each invariance type's checkers can help identify are indicated. There are invariance checker modules capable of localizing faults on their own. Invariance checker module 27 is missing, since it is only applicable to nonatomic VC buffers

Fig. 9.8 Probability of ambiguity in the output of the Localization Unit, when faults are localized at the port-level granularity

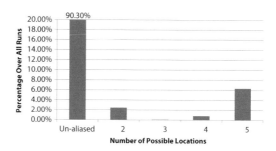

Localization Unit, when faults are localized at the port-level granularity. When the output is ambiguous (10%), the Localization Unit reports more than one (up to 5) possible faulty ports in the affected router, as shown in Fig. 9.8, with the faulty port always being reported.

9.3.4.3   RTL Netlist-Level Evaluation

As described in Sect. 9.3.3.1, a baseline NoC router augmented with the NoCAlert detection mechanism was implemented in Verilog HDL and synthesized using 65-nm commercial standard-cell libraries. The obtained netlist was used to evaluate NoCAlert's *detection* performance at the RTL netlist-level. The evaluation process at this level was kept the same as the corresponding high-level (network-level) methodology. Experiments were performed under a fault-free scenario to generate a Golden Reference, which was later compared with the output logs of the fault-injected simulation runs.

Similar to Fig. 9.4, Fig. 9.9 presents a breakdown of the RTL netlist-level fault-detection performance over two different fault-injection time periods: cycles 10–39 (empty NoC), and cycles 100–129 (warmed-up NoC). Note that the RTL netlist-level evaluation was conducted using only the transient fault model. Thus, we

Fig. 9.9 Breakdown of the RTL netlist-level fault-detection performance over two different fault-injection time periods

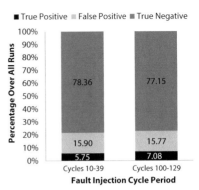

Fig. 9.10 Cumulative fault-detection delay distribution for True Positive faults, when simulating at the RTL netlist-level

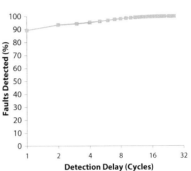

compare the RTL results of Fig. 9.9 with the corresponding "Transient" bars in Fig. 9.4. The first major observation is that **even under this extremely detailed evaluation, NoCAlert yields 0 % False Negatives**, thereby demonstrating full fault coverage.

Finally, Fig. 9.10 shows the cumulative fault-detection delay distribution for True Positive faults, and it is the RTL netlist-level equivalent of Fig. 9.5. The RTL delay trend is very similar to the network-level behavior, while the worst-case fault-detection latency is almost identical to the one reported at the network-level.

9.4 Conclusions

The advent of the multi-/many-core paradigm has elevated the criticality of the NoC. Being an integral part of the system, the NoC consumes non-negligible amounts of power. Additionally, the NoC's criticality—in the sense that it can become a single point of failure—increases the need for detecting and localizing possible faults.

The first work presented in this chapter has proposed a novel fine-grained power-gating methodology, aptly called *BlackOut*, which can operate at the granularity of individual VC buffers within each input port. The new mechanism entails minimal changes to the router architecture, and it is generally applicable to any pipeline micro-architecture, any topology, and any routing algorithm. The new methodology is fully distributed, and—unlike competing approaches—it only requires minimal information exchange between adjacent router pairs. The *BlackOut* technique relies on a brace of elemental micro-architectural concepts that operate in tandem: *late binding*, and *flow balancing*. Flow balancing enables smooth reactions within the power-gating actuator, which minimize the impact on performance. Late binding remaps new incoming packets to already-on buffers, thereby avoiding unnecessary wake-ups of "sleeping" buffers.

Extensive evaluations using both synthetic traffic and real application workloads validate the effectiveness of the *BlackOut* methodology. Compared to a baseline NoC, *BlackOut* can achieve energy savings of up to 70%, with a minimal 2% performance overhead. Most importantly, the new mechanism is compared to two state-of-the-art techniques, which represent the leading solutions in router- and buffer-level power-gating granularities. *BlackOut* is demonstrated to significantly outperform both techniques, by 35%, on average, in terms of energy savings.

The second work presented in this chapter proposes NoCAlert, a comprehensive **online and real-time fault *detection* and *localization*** mechanism that ensures 0% false negatives within the NoC, under both transient and permanent faults. NoCAlert is based on the concept of *invariance checking*, whereby the outputs of the control logic modules of the on-chip network are constantly checked for *illegal* outputs, based on current inputs. A collection of such micro-checker modules is used to implement real-time hardware assertions. The checkers operate concurrently with normal NoC operation, thus obviating the need for periodic, or trigger-based, self-testing. Upon fault detection, the status of the checkers within each router is analyzed by a Localization Unit, which can accurately pinpoint the fault location at three different supported granularity levels.

Extensive simulation results—both at the **network-level** and at the **RTL netlist-level**—validate the efficacy of the NoCAlert mechanism and yield important insight as to the behavior of the network when non-invariant faults (that evade the checkers) occur. Specifically, non-invariant faults either cause some subsequent invariance violation (and are captured), or they prove benign at the network/system level. Furthermore, a detailed comparison with a state-of-the-art framework [22] highlights higher than $100\times$ improvements in detection latency, with no loss in detection accuracy and with much lower overall complexity.

In summary, NoCAlert demonstrates the potential for near-instantaneous fault detection and swift, single-cycle fault localization within the NoC. This feat is achieved using fully distributed fault-localization units.

References

1. Agarwal, N., Krishna, T., Peh, L.-S., & Jha, N.K. (2009). GARNET: A detailed on-chip network model inside a full-system simulator. In *ISPASS*. https://doi.org/10.1109/ISPASS. 2009.4919636.
2. Binkert, N., Beckmann, B., Black, G., Reinhardt, S.K., Saidi, A., Basu, A., et al. (2011). The Gem5 simulator. *SIGARCH Computer Architecture News*. https://doi.org/10.1145/2024716. 2024718.
3. Borrione, D., Helmy, A., Pierre, L., & Schmaltz, J. (2007). A generic model for formally verifying NoC communication architectures: A case study. In *NOCS*. https://doi.org/10.1109/NOCS.2007.1.
4. Brayton, R. K., Sangiovanni-Vincentelli, A. L., McMullen, C. T., & Hachtel, G. D. (1984). *Logic minimization algorithms for VLSI synthesis*. Norwell: Kluwer Academic Publishers.
5. Chen, L., Zhu, D., Pedram, M., & Pinkston, T. M. (2015). Power punch: Towards non-blocking power-gating of NoC routers. In *HPCA*. https://doi.org/10.1109/HPCA.2015.7056048.
6. Chrysanthou, K., Englezakis, P., Prodromou, A., Panteli, A., Nicopoulos, C., Sazeides, Y., (2016). An online and real-time fault detection and localization mechanism for network-on-chip architectures. In *TACO*. https://doi.org/10.1145/2930670.
7. Constantinides, K., Plaza, S., Blome, J., Zhang, B., Bertacco, V., Mahlke, S., et al. (2006). BulletProof: A defect-tolerant CMP switch architecture. In *HPCA*. https://doi.org/10.1109/HPCA.2006.1598108.
8. Das, R., Narayanasamy, S., Satpathy, S. K., & Dreslinski, R. G. (2013). Catnap: Energy proportional multiple network-on-chip. In *ISCA*. https://doi.org/10.1145/2485922.2485950.
9. Fick, D., DeOrio, A., Hu, J., Bertacco, V., Blaauw, D., & Sylvester, D. (2009). Vicis: A reliable network for unreliable silicon. In *DAC* (pp. 812–817).
10. Galles, M. (1997). Spider: A high-speed network interconnect. *IEEE Micro, 17*, 34–39.
11. Hoskote, Y., Vangal, S., Singh, A., Borkar, N., & Borkar, S. (2007). A 5-GHz mesh interconnect for a teraflops processor. *IEEE Micro*. https://doi.org/10.1109/MM.2007.4378783.
12. Hosseinabady, M., Dalirsani, A., & Navabi, Z. (2007). Using the inter- and intra-switch regularity in NoC switch testing. In *DATE*. https://doi.org/10.1109/DATE.2007.364618.
13. Kakoee, M. R., Bertacco, V., Benini, L. (2011). A distributed and topology-agnostic approach for on-line NoC. In *International Symposium on NoCs*. https://doi.org/10.1145/1999946. 1999965.
14. Kakoee, M. R., Bertacco, V., Benini, L. (2011). ReliNoC: A reliable network for priority-based on-chip communication. In *DATE* (pp. 1–6).
15. Matsutani, H., Koibuchi, M., Ikebuchi, D., Usami, K. Nakamura, H. & Amano, H. (2011). Performance, area, and power evaluations of ultrafine-grained run-time power-gating routers for CMPs. In *TCAD*. https://doi.org/10.1109/TCAD.2011.2110470.
16. Meixner, A., Bauer, M. E., & Sorin, D. (2007). Argus: Low-cost, comprehensive error detection in simple cores. In *International Symposium on Microarchitecture*. https://doi.org/10.1109/MICRO.2007.8.
17. Mishra, A., Vijaykrishnan, N., & Das, C. (2011). A case for heterogeneous on-chip interconnects for CMPs. In *ISCA 2011* (pp. 389–399).
18. Moscibroda, T., & Mutlu, O. (2009). A case for bufferless routing in on-chip networks. In *ISCA '09* (pp. 196–207).
19. Nassif, S. R., Mehta, N., & Yu, C. (2010). A resilience roadmap. In *DATE* (pp. 1011–1016).
20. Nicopoulos, C., Srinivasan, S., Yanamandra, A., Dongkook, P., Narayanan, V., Das, C. R., et al. (2010). On the effects of process variation in network-on-chip architectures. *IEEE Transactions on Dependable and Secure Computing*. https://doi.org/10.1109/TDSC.2008.59.
21. Parikh, R., & Bertacco, V. (2014). ForEVeR: A complementary formal and runtime verification approach to correct NoC functionality. *ACM Transactions on Embedded Computing Systems*. https://doi.org/10.1145/2514871.

22. Parikh, R. & Bertacco, V. (2011). Formally enhanced runtime verification to ensure NoC functional correctness. In *MICRO*. https://doi.org/10.1145/2155620.2155668.
23. Prodromou, A., Panteli, A., Nicopoulos, C., & Sazeides, Y. (2012). NoCAlert: An on-line and real-time fault detection mechanism for network-on-chip architectures. In *MICRO*. https://doi.org/10.1109/MICRO.2012.15.
24. Strano, A., Gómez, C., Ludovici, D., & Favalli, M., Gómez, M. E., & Bertozzi, D. (2011). Exploiting network-on-chip structural redundancy for a cooperative and scalable built-in self-test architecture. In *DATE* (pp. 1–6).
25. Sun, C., Chen, C. H. O., Kurian, G., Wei, L., Miller, J., Agarwal, A., et al. (2012). DSENT - A tool connecting emerging photonics with electronics for opto-electronic networks-on-chip modeling. In *NoCS*. https://doi.org/10.1109/NOCS.2012.31.
26. Woo, S. C., Ohara, M., Torrie, E., Singh, J. P., & Gupta, A. (1995). The SPLASH-2 programs: Characterization and methodological considerations. In *ISCA* (pp. 24–36).
27. Zoni, D., Canidio, A., Fornaciari, W., Englezakis, P., Nicopoulos, C., & Sazeides, Y. (2017). BlackOut: Enabling fine-grained power gating of buffers in network-on-chip routers. *Journal of Parallel and Distributed Computing*. https://doi.org/10.1016/j.jpdc.2017.01.016.
28. Zoni, D., Flich, J., & Fornaciari, W. (2016). CUTBUF: Buffer management and router design for traffic mixing in VNET-based NoCs. *IEEE Transactions on Parallel and Distributed Systems, 27*(6), 1603–1616. https://doi.org/10.1109/TPDS.2015.2468716.
29. Zoni, D., & Fornaciari, W. (2013). Sensor-wise methodology to face NBTI stress of NoC buffers. In *DATE*. https://doi.org/10.7873/DATE.2013.216.
30. Zoni, D., & Fornaciari, W. (2015). Modeling DVFS and power-gating actuators for cycle-accurate NoC-based simulators. *ACM Journal on Emerging Technologies in Computing Systems, 12*, 1–15. https://doi.org/10.1145/2751561.
31. Zoni, D., Terraneo, F., & Fornaciari, W. (2015). A DVFS cycle accurate simulation framework with asynchronous NoC design for power-performance optimizations. *Journal of Signal Processing Systems*. https://doi.org/10.1007/s11265-015-0989-1.

Part V
Technology Related Reliability Approach

Chapter 10
Time-Efficient Modeling and Simulation of True Workload Dependency for BTI-Induced Degradation in Processor-Level Platform Specifications

Simone Corbetta, Pieter Weckx, Dimitrios Rodopoulos, Dimitrios Stamoulis, and Francky Catthoor

10.1 Workload-Dependent Platform/Processor-Level Analysis of Bias Temperature Instability-Induced Degradation: Goal and Motivation

Bias temperature instability (BTI) is important for digital system reliability, and EDA techniques have been proposed to model and mitigate the BTI-induced variability [14]. However, the aggressive scaling toward deca-nanometer nodes has significantly changed the modeling perception of BTI and the underlying simulation requirements. Recent literature shows that older models inherently fail to capture BTI phenomena at deeply scaled technology nodes [7]. Previously, an atomistic approach has been proposed which concentrates on the charge-level activity of gate-stack defects [9]. It attributes the manifestation of BTI to a variable number of defects, each one with a specific temporal behavior (modeled using time constants) and a certain contribution to the overall V_{th} degradation of a transistor. Recent EDA literature supports a paradigm shift to defect-centric modeling [7] and a variety of atomistic implementations already capture BTI-induced degradation over arbitrary circuit lifetime intervals [23], or in a transient way. More recent approaches are based on the analytical modeling of defect activity using capture/emission time (CET) maps [6]. CET maps describe the probability density function of defect capture and emission times as well as their correlations at high and low digital

S. Corbetta (✉) · P. Weckx · D. Stamoulis · F. Catthoor
Imec, Leuven, Belgium
e-mail: simone.corbetta@polimi.it; pieter.weckx@imec.be; dimitrios.stamoulis@mail.mcgill.ca; francky.catthoor@imec.be

D. Rodopoulos
MicroLab-ECE-NTUA, Athens, Greece
e-mail: drodo@microlab.ntua.gr

© Springer International Publishing AG, part of Springer Nature 2019
W. Fornaciari, D. Soudris (eds.), *Harnessing Performance Variability in Embedded and High-performance Many/Multi-core Platforms*,
https://doi.org/10.1007/978-3-319-91962-1_10

operating voltages [17]. However, monitoring the activity of several defects per transistor throughout the design lifetime poses severe restrictions on the circuit size, the amount of time samples (and hence the granularity of the workload stimuli), and simulation time. Consequently, existing works limit their exploration to simple gates [10, 12], SRAM cells [23, 24], and simple logic blocks [13, 21].

Our goal is to employ the atomistic CET-based models for variability analysis along with the efficiency of compact digital waveform (CDW) approximation to efficiently capture the time-varying BTI-induced path delay degradation on the true workload, i.e., without assuming distributed or constant profiles of the stress load. In this chapter, we adapt the analytical CET map modeling method to the platform level, in such a way that we are able to achieve accurate yet efficient atomistic workload-dependent variability analysis for potentially any reference design and desired workload. We then provide examples which demonstrate this simulation capability on realistic CPU components and benchmark applications.

10.2 Technology-Based Modeling of Platform Performance Degradation

Performance degradation is the combined effect of different physical mechanisms that result in the divergence of sensible quantities (electrical metrics for devices and performance metrics for circuits) from their nominal expected boundaries. The very source of such phenomena is out of the scope of this discussion, and the interested reader can refer to previous deliverables [18] and [11]. This section will instead briefly report the interconnection between the low-level models and the high-level perspective in HARPA, specifying the course of action required to feed upper levels of the computing stack.

10.2.1 Workload Dependency

It is well known that the induced aging is a function of electrical parameters (supply voltage), environmental variables (temperature), performance requirements (frequency), and workload [2]. This can be captured in a variety of ways, depending on the focus of the analysis, and on the information that is available about the reference circuit. In general, system-level analysis is required to estimate the impact of reliability phenomena to the user. However, the necessary information is not always readily available for the purpose: in case of complex commercial processors, indeed, the design database is of typically not publicly accessible, while the use of real hardware poses severe constraints due to the black box approach required to gather low-level information. In either case, in the HARPA approach we have addressed the workload dependency issue, as discussed next.

10.2.2 Black Box Approach for System-Level Analysis

As soon as the design is closed and no precise information can be gathered, the design is basically a black box: only inputs and outputs are known. It is however important to cope with these situations, as many commercial platforms fall under this category. The amount of information that can potentially be retrieved here is still relatively high, but the abstraction level is also high. Within HARPA a methodology has been developed to analyze these systems, and this is reported in Chap. 11.

10.2.3 White Box Approach for Partitioned Processor-Level Analysis

When the internals of the target design are known, and a detailed platform netlist is hence available, a white box approach can be adopted. For our purposes, the design and verification database contains then all the information which is required. In this best case scenario, all the constituents of the system can be efficiently and accurately analyzed.

Let us assume for a moment the high-level block diagram of a generic processor as illustrated in Fig. 10.1. All main blocks are reported here, including digital, memory, and analogue components. Each of these types of components requires a separate methodology. For memory blocks (as present in the caches, for example), we have developed a specific way of workload-dependent modeling and analysis which is summarized in [1, 24]. That work is not the focus of this chapter. And, for analogue components, the devices are typically quite large which implies that the strong BTI effects due to scaling are largely avoided there. So, that is also not our focus here. Instead, we are dealing with the digital logic processor cores. And also, these can be largely analyzed as separate pipeline stages when focusing on parametric performance degradation. The register files or register clusters and

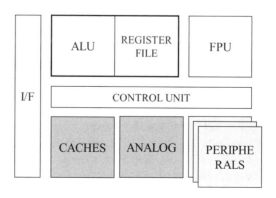

Fig. 10.1 Partitioning of the reference design into large isolated blocks with different characteristics

the neighboring logic components like ALUs, multipliers, and FPUs (floating-point units) will then still be combined in a full pipeline stage.

The processor-level analysis can thus be designed to efficiently provide performance degradation breakdown on a per-module basis. In general, to cope with the complexity of the circuit and to filter out non-digital logic modules (e.g., caches and analogue plane), the design is tore down into smaller components; these are isolated at first hand, dependencies (i.e., connection information) are solved and maintained up-front and analysis can then be performed efficiently. It is worthwhile to notice that it is up to the design engineer at this point to select the most appropriate partitioning scheme, bearing in mind that this is context-dependent and a fine-versus coarse-grain partitioning study should be conducted. This decision directly impacts the way performance degradation metrics are computed.

With this approach, all the inputs and outputs are known, but all the internal nodes are known as well. This approach gives lots of flexibility, but usually at higher costs in terms of design time to be spent. It will be further detailed in the rest of this chapter.

First, in Sect. 10.1 the goal will be defined, followed by the related work in Sect. 10.3. Next, we will outline the different steps of our approach in Sect. 10.4. And then, we will show the main results obtained in Sect. 10.5.

10.3 Related Work at the Component or Platform Level

To the best of our knowledge, this is the first scalable approach that employs complete atomistic variability analysis of realistic CPU subsystems, while retaining the true workload dependency under different reference applications. A large body of previous works has focused on either simple CMOS logic gates [10, 12], SRAMs [20, 23, 24], or larger netlists of repetitive blocks with reduced functional complexity [4, 21]. Rodopoulos et al. have implemented atomistic models on top of transient SPICE simulators [19, 21]. Although significant accuracy is achieved, the use of SPICE-based solvers is not scalable to larger circuits, resulting in prolonged execution times up to 100 h even for a simple circuit with few gates. Weckx et al. have enabled atomistic analysis for SRAM cells with numerical methods [23, 24]. Nonetheless, the formulation is bound to the given six-transistor topology and may be complicated to extend to arbitrary topologies with more transistors. Consequently, these approaches are not applicable to larger digital logic netlists due to scalability issues. Moreover, only two existing works enable atomistic BTI analysis at the subsystem level, similarly to our approach. Kukner et al. have applied CET map modeling methodologies to enable workload-dependent analysis for several logic blocks [13]. Nonetheless, the evaluation is limited to only restricted synthetic workloads. Relatively short varying patterns are randomly assigned to the input pins of each block and such activity information is propagated to the inner nodes. This does not enable true workload-dependent reliability analysis. Instead, in this chapter we demonstrate the usability of our approach for several large realistic test cases with realistic deterministic workloads. Stamoulis et al. have recently enabled pseudo-transient atomistic simulations for an entire CPU datapath [22]. Nevertheless, the impact of workload dependence is completely neglected and the same stress pattern is assumed for all the transistors, regardless of their relative position in the netlist and the activity factors of the inner nodes. Such an assumption is questionable, since these signal attributes are not constant in modern complex circuits, and poses a limitation to the desired BTI estimation. In Sect. 10.5, we show that such assumption leads to fundamental estimation errors that cannot be neglected in future design for reliability in scaled technology nodes. In [5], the authors proposed a statistical analysis framework to obtain the combined effect of process variation and NBTI on the delay distribution; however, they did not consider the intrinsic variability of NBTI. The proposed gate-level simulation flow in [23] uses a look-up table (LUT)-based approach where each cell is characterized for different combination of threshold voltage shift (V_{th}) of its transistors and the timing information is stored in LUTs in an extended library. Stochastic samples are interpolated from the LUTs used by timing analysis tool to obtain the timing information of the gates.

In addition, industrially oriented commercial EDA flows exist. Accurately assessing the full impact of the stochastic behavior of BTI is typically only performed at the device level as the circuit-level timing analysis poses severe restrictions on the circuit size and/or the simulation time. In order to reduce the complexity, some flows have been developed which exploit circuit-level statistical timing analysis on the delay distribution (STA). These frameworks currently consider only a single critical path [3] or the analysis is only performed at the $+3\sigma$ corner [20].

10.4   Bias Temperature Instability-Induced Degradation Analysis Flow

The EDA flow to fulfill the white box approach objectives introduced in Sect. 10.2.3 is now presented in more detail. The proposed approach enables variability analysis that retains both the atomistic property [23] and workload memory [20].

As shown in Fig. 10.2, the flow is meant to be employed at an early stage of the design process, facilitating BTI-aware design of digital circuits. Design engineers will be able to perform early analysis of performance variability to foster eventual design choices. As soon as the gate-level netlist from synthesis is available, iterations are possible, and design updates can be performed. In case parts of it are missing, as in the HARPA case, alternative strategies to retrieve estimates can be adopted, as already explained in Sect. 10.2.2. The flow is designed to help digital and integration engineers in revealing potential reliability flows in their design, considering the target workload and reference operating conditions. This happens early at the design stage, so that design refinement can happen before even entering the implementation phase, at which point any change would be really costly.

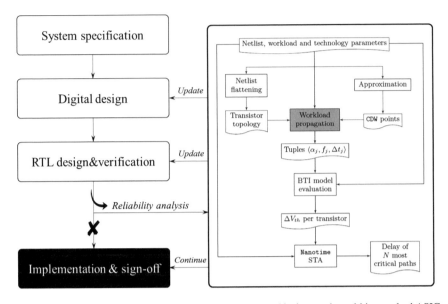

Fig. 10.2 High-level representation of the proposed flow, and its integration within standard ASIC digital design flow

The proposed flow is based on standard and commercial ASIC design tools integrated with in-house control ones; however, implementation details are out of the scope of this document. The very last outcome is the set of critical paths in the design before and after the aging analysis has been applied. This gives information about those portions of the design that are subject the most to aging (as a consequence of the applied workload). They represent the most important contributions to the performance degradation, and such information can be used to define a protection strategy at the digital or RTL design level. As soon as aging meets the expectations, the implementation phase can then take place. Notice that more accurate information from the back-end phase could be used to drive the aging analysis, but this would increase the turnaround time, thus failing in providing early reliability insights.

The main steps in our analysis methodology for performance degradation estimation will be presented in more detail in the next subsections.

10.4.1 Netlist Flattening

The BTI analysis has to be performed at the transistor level, since the atomistic and probabilistic properties of the BTI theory [5, 6, 23] are based on the profile of the stressing V_{gs} voltage. For this purpose, the gate-level circuit description (from the synthesis phase) is translated into transistor-level using the gates description available in the technology library database. A renaming script is used to derive consistently connected network of transistors that describe the reference circuit (this step is called transistor topology in Fig. 10.2). Notice that this process causes the number of primitives to analyze to explode, since (at least) a factor of 4 has to be assumed while moving from CMOS gates to transistors count. This phase is also important to determine the list of all net names that need to be taken care of in the subsequent phases.

10.4.2 Workload Approximation

Monitoring the occupancy of several defects per transistor in a transient fashion imposes severe restrictions on scalability [10]. We employ an accurate yet efficient method to capture the evolution of stress patterns throughout the circuit lifetime. In previous work, a novel signal waveform representation has been proposed, namely CDW [20]. The analysis of the reference workload is reduced to a set of hyphenated CDW points; they represent the regions of a signal in which the impact of the workload on the BTI-induced variation will be computed. For each region with duration Δt, the CDW point is expressed by means of the frequency f and duty factor α of the stress signal across the region.

The user is able to determine the way the waveform is approximated in either one of two ways: by specifying the number of final events (associated with CDW points) or by specifying the duty cycle and frequency margins that the approximation algorithm will use. In the former case, the use is basically dividing the waveform into a number of intervals with uniform length. In the latter case, on the other hand, an enhanced version of the algorithm presented in [20] is employed.

In general, the choice of the number and position of these points can play an important role, since it determines the accuracy of the estimation and the simulation overhead. In general, a higher number of CDW points make the analysis more accurate at the cost of simulation time, while a more crude analysis suffers from averaging out the signal activity. This trade-off analysis is beyond the scope of the current chapter, and the interested reader is referred to the literature [20]. Nevertheless, we can exploit the proposed flow to estimate the relative divergence of the BTI estimation as a function of the number of selected CDW points.

In practice, the reference workload is described through the VCD (Value Change Dump) standard file format, a convenient way to describe digital waveforms in text format (easy to parse and process). The VCD file is a list of time/value pairs for a generic signal: an entry represents an event with the absolute time at which a signal changes and its new value. For its structure, VCD files can contain several millions of events for longer workloads. For complex circuits with many signals, this file becomes huge, and the number of events are not manageable. We then exploit the already explained CDW concept. In this way, the original sequence is approximated with a much shorter number of events that describe the former workload with a

similar one. The resulting waveform will have on average the same characteristics of
the original one; the number of events in the final sequence determines the similarity
of the two waveforms.

10.4.3 Workload Propagation

In order to retain the realistic workload dependency across all the nodes of the
circuit, the stress activity patters need to be propagated throughout all the transistors
of the digital block in terms of the tuples containing the duty factor α, the frequency
f, and the duration Δt. Furthermore, we need to cope with the space requirements
for complex circuits and large waveforms; tracing activity factors for all the internal
nodes from RTL simulation would be unmanageable.

For this reason, we collect complete waveforms at the primary inputs of the
desired circuit abstraction, and flatten the standard cells of the gate-level design
to their equivalent transistor topology (based on their SPICE-level descriptions
from the reference technology library). We then automatically translate it to a
second Verilog circuit description that is topologically equivalent to the SPICE-
level one, but that can be still simulated using digital simulation tools. The process is
illustrated in Fig. 10.3. Based on this, we then perform switch-level simulation over
consecutive CDW points using Cadence NCSIM, and compute the signal activities
at each transistor from activity TCF data instead of keeping complete VCD waves.
Switch-level simulation is configured so that the top-level stimuli are propagated to
the transistor-level topology description, and TCF (toggle count format) information
is generated for the V_{gs} of each transistor. Based on the resulting activity files for
each region with duration Δt, we can easily compute the α and f values at each
transistor. The resulting CDW representation captures the V_{gs} stress voltage at the
transistor level for all the nodes of the netlist under test, as required by the CET
maps model. Notice that V_{gs} is the BTI stress waveform, and it differs conceptually
from the digital waveform.

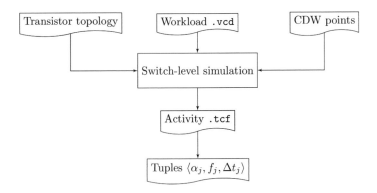

Fig. 10.3 Propagation flow based on switch-level simulation

10.4.4   Workload-Dependent Aging Computation

Once the stress patterns of each transistor are known, we employ atomistic BTI modeling over all the CDW points to accurately evaluate the BTI-induced degradation of each transistor in terms of threshold voltage shift ΔV_{th}. By combining the features of both the CDW and CET-map methodologies for representative workloads on full processor components, we then formulate our BTI evaluation based on the model presented in [23]. For each trap of each transistor, the occupancy probability is evaluated, depending on the applied stress waveform specified by the current CDW point, to compute the contribution to the overall ΔV_{th} shift.

We repeat this process over all the CDW points to account for workload and time dependency. Notice that each transistor has to be analyzed in a separate way due to the stochastic nature of the problem [23]. In this way, we derive the workload-dependent ΔV_{th} per transistor. Finally, by employing a transistor-level STA tool, namely Synopsys NanoTime, we compute the BTI-aware delay values, keeping track of its evolution throughout the duration of the reference application.

10.4.5   Circuit-Level Aging Analysis

The last step is to translate the per-transistor threshold voltage shift into a circuit-level degradation metric in order for the aging impact to be of any use and meaning. Transistor-level static timing analysis is then performed. During the analysis, each transistor in the circuit has a threshold voltage value that is the result of the aging stress computed during the previous phase. The output of the STA process is an ordered list of critical (and the so-called near-critical) paths that result from the workload-induced aging. Comparison of this list with the non-aged one (i.e., transistors with their nominal threshold voltage) will reveal the aging effects on the paths timing and, ultimately, on the circuit performance.

It is important to notice here that the analysis can be performed in two ways: expansive, based on Monte Carlo sampling, or worst case based on $n - \sigma$ analysis. In the former case, a Monte Carlo is added on top of the aging process reported in Sect. 10.4.4, where the stochastic process is the traps emission/capture probability determined by theoretical model equations [23]. This sampling will give a distribution of the path delays, while the $n - \sigma$ will give a single value. Usually, we use $3 - \sigma$ analysis, but $6 - \sigma$ analysis can be performed as well, if worst case has to be pushed further, as expected for lower technology nodes.

10.5 Representative Results

The purpose of this section is to show representative results while performing BTI analysis with the overall HARPA methodology explained so far. We substantiate the validity of our ow as a useful aid for early stage BTI analysis and BTI-aware design techniques, by capturing the workload-dependent delay degradation for selected CPU modules. We provide an analysis of a few reference circuits taken from a representative commercial design. During our evaluation, the workload is generated through RTL simulation using simple, but real-world applications. An extensive experimental campaign has been conducted to capture workload dependency of BTI-induced aging in an accurate way and reasonable turnaround time.

In the following subsections, the integer add/sub component will be used as main test vehicle to discuss on results, and few additional results will be shown for other components as well. The electrical and physical parameters employed in the transistor-level analysis are then taken directly from the model card of bulk CMOS nodes of TSMC technology. Calibration has been performed against available silicon measurements [16].

10.5.1 Experimental Setup

10.5.1.1 Target Technology

We have employed the standard-cell library and device models from the TSMC 40 nm technology; the calibration of the flow technology-dependent parameters has been conducted using a simple inverter gate, and considering a reference degradation at reference operating conditions. However, the flow is designed in such a way that any technology library can be used, provided that the model-card file and the subcircuits file are provided. Thus, calibration can be performed as well, even for different operating conditions (subject to the availability of published silicon technology data).

10.5.1.2 Reference Circuits

For our experiments, we selected a number of digital subcircuits from the OpenSPARC T1 processor, whose design and verification environment is freely available [15]. This processor is representative of a complex and commercial full-chip design. We have focused our analysis on selected modules from the integer and floating-point (IEEE- 754 2008 double-precision standard compliant) datapath, ranging from 6k equivalent gates (or 22k transistors) of the integer ALU to 36k equivalent gates (or 143k transistors) of the floating-point multiplier. These are reported in Table 10.1. For each of them, the number of equivalent gates and the grand total number of transistors is listed. The number of equivalent gates is

Table 10.1 Selected circuits from the OpenSPARC T1 design

Module	Number of equivalent gates	Number of transistors
Integer ALU	6k	22k
Integer multiplier	34k	135k
Floating-point ALU	29k	115
Floating-point multiplier	36k	143k
Floating-point divider	11k	44k

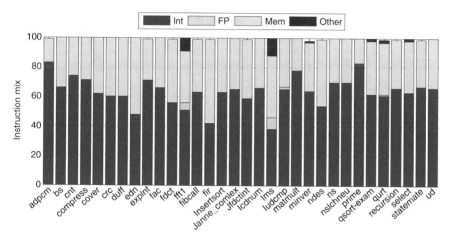

Fig. 10.4 WCET instructions breakdown

computed using the simplest 2-input NAND gate available in the TSMC 40 nm technology library. The circuits have been isolated during synthesis, so that a top level with no hierarchy for each of them is available.

10.5.1.3   Reference Applications

A number of applications from the WCET benchmark suite [8] have been selected to cover a wide range of application characteristics. The breakdown of instructions from the WCET suite is given in Fig. 10.4.

Unless otherwise noted, the waveforms from the above applications are expanded to years of operation: instead of extending the inputs to cover longer simulation times, the original short waveforms are expanded (across CDW points) in such a way that the sum of the simulated time is years. This is expressed in the example of Fig. 10.5. Let us assume that a single waveform that lasts for 300 ms is available, and let us assume that the user imposed 6 CDW points, such that each BTI simulation accounts for 50 ms of the total time. The expansion process takes these CDW intervals and expand them in such a way that (for a total simulated time of 6 months) they will account for 1 month each. The example reports just one single CDW

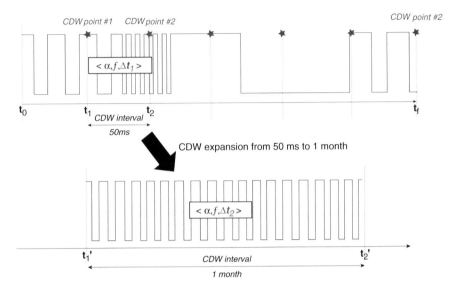

Fig. 10.5 Compact digital waveform (CDW) expansion

interval, but the same process is repeated for the others as well. It is the duration time parameter Δt of the tuples that changes, while the duty cycle and frequency do not, retaining the waveform characteristics and mimicking long stress waveforms.

10.5.2 Constant Versus Realistic Activity Factors

In this subsection, we demonstrate that an invalid assumption about the stress load profile to PMOS and NMOS transistors can lead to an erroneous degradation estimation. We compare our results that retain workload-dependent attribute against the results with the state-of-the-art proposal in [22], in which the authors assumed a constant activity factor and frequency across all transistors. We then adapted our flow to resemble this experimental setup, assuming a 50% duty factor and 1 GHz frequency at each transistor and one single CDW point (that is placed exactly at the end of the workload). In this way, each transistor will be subject to the same exact stress, irrespective of the application generating the workload.

We perform variability analysis for an overall stress duration of 3 years, as in [22]. To perform this analysis, we started from the CDW points computed by the ow for a number of applications, and we then scaled the duration of each CDW interval in such a way that the sum of all of them equals the desired wall-clock time. In each case, we maintained the original ratio among each CDW intervals. For the integer add/sub circuit, we compare the ΔV_{th} at each transistor, computed based on the constant duty-factor case and the simulated one in Fig. 10.6. It shows

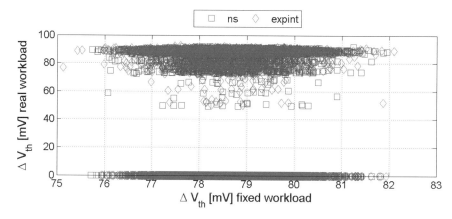

Fig. 10.6 Dispersion plot comparing data samples from constant duty-factor case and simulated cases

the dispersion plot of the ΔV_{th} values in two different applications: each point represents the threshold voltage shift at the end of the workload of each single transistor. The scatterplots highlight the mismatch of homonym transistors in the two scenarios. Two clouds are evident. First of all, the topmost cloud expresses a mismatch due to the assumed workload even for alternate stresses. On the other hand, the bottom cloud highlights the presence of idle transistors in the circuit. These transistors will be however computed as degrading assuming a constant factor as in [22], thus providing a fundamental estimation error. Samples are just reported for two different applications (ns and expint), but similar results have been obtained for the other applications as well.

The importance of the mismatch can also be evaluated by comparing the empirical cumulative distribution function (CDF) of the real stress case with constant stress case (i.e., 10 and 90%) and a wider time range (i.e., 1, 3, and 5 years of operation), for three different applications matmult, fir, and compress. The empirical CDF F(x) of the collected samples ΔV_{th} are shown in Fig. 10.7. Lines are grouped according to the scenario, either of a constant workload or of a simulated one. The analyzed wall-clock time values are denoted by the curves shifting toward the right. The results show that the low (10%) and the high (90%) constant duty-factor (unchanged frequency) assumptions enclose the real aging evolution up to a certain degree. Beyond a certain interval, the estimation is basically erroneous. Also, the shape of the functions differs from the real case. It is worth observing that a higher stress time moves the curves toward the right-end side of the axis. For instance, assuming a ΔV_{th} degradation equal to 80 mV, the 10% assumption would over-constrain the design, since the cumulative probability at this point has already reached 1; for the 90% case, on the other hand, the design would be under-constrained, even for 12 months reference time. It is therefore evident that we cannot avoid the workload dependency analysis if we are to deliver cost-effective design solutions.

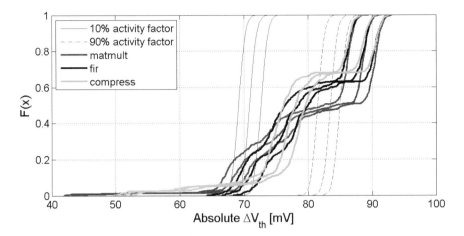

Fig. 10.7 Empirical CDF shows divergence between constant and real workload

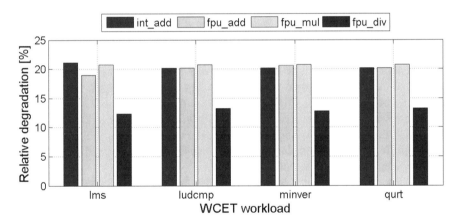

Fig. 10.8 Workload/circuit pairs comparison

10.5.3 Impact of Workload

Figure 10.8 reports the relative degradation (in percentage with respect to the base case with no aging) of different circuits, considering different common workloads. The results are given after a simulated time (expanded as expressed in Sect. 10.5.1) of 3 years at 125 °C and nominal voltage of 1.0 V. The chosen applications are floating-point intensive ones, meaning that they contain a number of floating-point operations for which the floating-point hardware unit is used. However, it is shown that also the integer ALU is subject to a degradation that is comparable to the one for the FPU. This is because integer operations are always in place in the application, even when it is a floating-point intensive one. These results show that there exists a

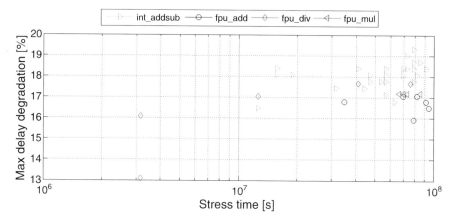

Fig. 10.9 Maximum degradation comparison

small degradation difference in the same circuit due to different applications. These results are in line with what is shown in the literature [4], namely that variability is reduced from device to circuit level, due to averaging effects in subsequent stages. This will be however investigated further by looking at specific test cases.

However, workload dependency is visible if we plot the maximum relative degradation against stress time for different cases, as in Fig. 10.9.

The time on the horizontal axis is the time required to reach the maximum degradation on the vertical axis. Different colors identify different circuits. The distribution of the sample points demonstrates that different workload/circuit pairs have a different aging profile. Orders of magnitude of difference in the stress time show the importance of workload on aging.

Figure 10.10 compares the time-dependent degradation of different workload/circuit pairs over a period of 5 years. The time-dependent degradation shows the impact of workload phases on the delay degradation of the most critical path.

10.5.4   Simulation Time Discussion

The time required to provide the BTI estimate depends on a number of factors, such as the number of transistors in the reference circuit, the topology of the circuit, and the length of the waveform (in number of events).

Figure 10.11 reports the cumulative estimation time for different workload/circuit pairs. The simulation time is divided into a number of phases, whose main contribution is specified in Table 10.2.

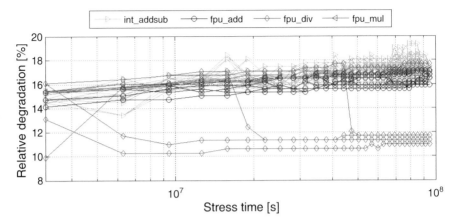

Fig. 10.10 Time-dependent degradation comparison

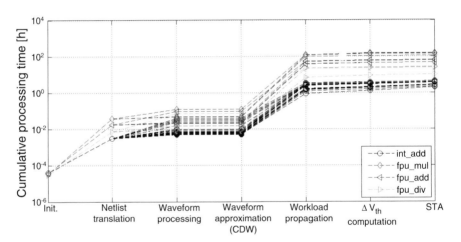

Fig. 10.11 Execution time comparison

10.5.5 Compact Digital Waveform-Based Approximation

For the flow presented in Sect. 10.4, the choice of the number of CDW points will have an impact on the final estimation. As a matter of fact, depending on the circuit and workload characteristics (that cannot be foreseen), the final BTI-induced degradation can give different results, especially in cases when the time-varying profile is of interest to the user. We provide a simple example in Fig. 10.12.

A simple test has been conducted using the integer ALU circuit with a fixed workload; the number of CDW points has been varied from 1 (a single BTI estimation covering the entire simulation time) and 100. The time-varying BTI-induced aging of different cases has then been overlapped in the figure. For the

Table 10.2 Phase contributions

Phase	Role	Main contributions
Netlist translation	Translates gate-level description to transistor-level description	– Netlist size (nr of gate instances) – Standard-cell library size (nr of available gates)
Waveform processing	Generates bit-level signals from port-wide signals of VCD file	– Circuit complexity (nr and width of signals) – Size of the waveform (nr of events in each signal)
Waveform approximation (CDW)	Computes the CDW approximation	– Size of the waveforms (nr of events in each signal)
Workload propagation	Propagates the workload to gate level	– Netlist size (nr of transistors) – Size of the waveform (nr of events in each signal)
ΔV_{th} computation	Computes BTI-induced threshold voltage shift	– Netlist size (nr of transistors)
Static timing analysis (STA)	Performs STA at transistor level	– Circuit complexity (nr of paths and topology) – Netlist size (nr of transistors)

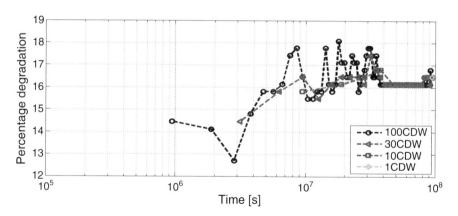

Fig. 10.12 Impact of number of CDW points on the BTI analysis

presented case, the last CDW point is comparable in all the different cases, but it is important to note that the time-varying profile is very different. Thus, the number of CDW points determines both the observability (in time) of the BTI profile. When the initial BTI impact has to be determined, it is important to choose a larger number of CDW points, and for cases in which the time-varying profile is of interest as well. For faster cases, on the other hand, a lower number of CDW points are enough.

However, the error introduced by using a lower number of CDW points can amplify with different circuit/workload pairs.

10.6   Conclusion and Future Work

In this chapter, we have proposed a comprehensive system-level reliability analysis flow that exploits the accuracy of CET maps with the efficiency of the CDW signal representation. Prior art on atomistic BTI modeling exhibits reduced scalability or accuracy issues, by either limiting the models to the circuit level or by assuming constant stress patterns at the system level. To the best of our knowledge, this is the first automated approach that retains both the atomistic property and true workload-dependent aging attribute.

That way, we enable variability analysis for several CPU subsystems, showing the importance of workload in BTI-induced aging analysis. First, we prove that worst-case margin-based proposals assuming constant workload (i.e., activity factor and frequency of the stress signal independent of the application) lead to over- or under-constrained designs. Moreover, we assess the impact of true workload dependency on different circuits, since the same applications result in clearly distinct overall degradation among different CPU modules. Thus, we fully demonstrate the usability of our tool flow for efficiently capturing the workload-dependent BTI-induced degradation of CPU components under realistic conditions for real-life applications.

So far, results have been presented for representative but still relatively limited use cases. However, our research work will continue toward considering even more complex scenarios. We are currently active in two main topics: flow optimization and gate-level analysis flow where the details of the transistor circuit inside complex gates can be abstracted away.

References

1. Agbo, I., Taouil, M., Kraak, D., Hamdioui, S., Kukner, H., Weckx, P., et al. (2017). Integral impact of BTI, PVT-variation and workload on SRAM sense amplifier. *IEEE Transactions on VLSI Systems, 25*(4), 1444–1454 (Online Jan 2017).
2. Catthoor, F., & Corbetta, S. (2014). *Harpa project deliverable D4.1: Report on high-level reliability model requirement specifications.* http://www.harpa-project.eu/index.php?option=com_content&view=article&id=10&Itemid=111&download=D41.pdf
3. Catthoor, F., Corbetta, S., & Meeus, W. (2015). *Harpa project deliverable D4.3: Proof-of-concept of HARPA modelling framework.* http://www.harpa-project.eu/index.php?option=com_content&view=article&id=10&Itemid=111&download=D43.pdf
4. Fang, J., & Sapatnekar, S. (2012, April). Understanding the impact of transistor-level BTI variability. In *IEEE IRPS Conference* (pp. CR.2.1–CR.2.6).
5. Grasser, T. (2014). The capture/emission time map approach to the bias temperature instability. In *Bias temperature instability for devices and circuits*. New York: Springer.

6. Grasser, T., Wagner, P. J., Reisinger, H., Aichinger, T., Pobegen, G., Nelhiebel, M., et al. (2011, December). Analytic modeling of the bias temperature instability using capture/emission time maps. In *IEEE IEDM Conference* (pp. 27.4.1–27.4.4).
7. Grasser, T., Kaczer, B., Goes, W., Reisinger, H., Aichinger, T., Hehenberger, P., et al. (2011). The paradigm shift in understanding the bias temperature instability: From reaction-diffusion to witching oxide traps. *IEEE Transactions on Electron Devices, 58*(11), 3652–3666.
8. Gustafsson, J., et al. (2010). The malardalen WCET benchmarks: Past, present and future. In *WCET 2010*, Brussels (pp. 136–146).
9. Kaczer, B., et al. (2010, May). Origin of NBTI variability in deeply scaled pFETs. In *IEEE IRPS Conference* (pp. 26–32).
10. Kaczer, B., et al. (2011, April). Atomistic approach to variability of bias-temperature instability in circuit simulations. In *IEEE IRPS Conference* (pp. XT.3.1–XT.3.5).
11. Kleanthous, M., Klokkaris, G., Nikolau, P., Chrysanthou, K., Hadjilambrou, Z., Englezakis, P., et al. (2014). *Harpa project deliverable D3.1 - Technological Context and State-of-the-Art.* http://www.harpa-project.eu/index.php?option=com_content&view=article&id=10&Itemid=111&download=D31.pdf
12. Kukner, H., et al. (2014). Comparison of reaction-diffusion and atomistic trap-based BTI models for logic gates. *IEEE Transaction on Device and Materials Reliability, 14*(1), 182–193.
13. Kukner, H., et al. (2014, March). Degradation analysis of datapath logic subblocks under NBTI aging in FinFET technology. In *15th ISQED Conference* (pp. 473–479).
14. Kuuoglu, H., & Alam, M. (2007). A generalized reaction-diffusion model with explicit h-h2 dynamics for negative-bias temperature-instability (NBTI) degradation. *IEEE Transaction on Electron Devices, 54*(5), 1101–1107.
15. OpenSPARC, Oracle. http://www.oracle.com
16. Reis, R., Cao, Y., & Wirth, G. (Eds.). (2015). *Circuit design for reliability.* New York: Springer.
17. Reisinger, H., et al. (2010, May). The statistical analysis of individual defects constituting NBIT and its implications for modeling DC- and AC-stress. In *IEEE IRPS Conference* (pp. 7–15).
18. Rodopoulos, D., & Soudris, D. (2014). *Harpa project deliverable D2.1 - State-of-the-art RT Mitigation of Performance Variability.* http://www.harpa-project.eu/index.php?option=com_content&view=article&id=10&Itemid=111&download=D21.pdf
19. Rodopoulos, D., et al. (2011, May) Time and workload dependent device variability in circuit simulations. In *IEEE ICICDT Conference* (pp. 1–4).
20. Rodopoulos, D., Weckx, P., Noltsis, M., Catthoor, F., & Soudris, D. (2014). Atomistic pseudo-transient BTI simulation with inherent workload memory. *IEEE Transaction on Device and Materials Reliability, 14*(2), 704–714.
21. Rodopoulos, D., et al. (2014, May). Understanding timing impact of BTI/RTN with massively threaded atomistic transient simulations. In *IEEE ICICDT Conference* (pp. 1–4).
22. Stamoulis, D., et al. (2015, May). Efficient reliability analysis of processor datapath using atomistic BTI variability models. In *25th ACM GLSVLSI Conference*.
23. Weckx, P., Kaczer, B., Toledano-Loque, M., Grasser, T., Roussel, P. J., Kukner, H. (2013, April) Defect-based methodology for workload-dependent circuit lifetime projections - Application to SRAM. In *IEEE IRPS Conference* (pp. 3A.4.1–3A.4.7).
24. Weckx, P., et al. (2014) Non-Monte-Carlo methodology for high-sigma simulations of circuits under workload-dependent BTI degradation - Application to 6T SRAM. In *IEEE IRPS Conference* (pp. 5D.2.1–5D.2.6).

Chapter 11
Proof-of-Concept HARPA Measurement-Based Platform Modelling Framework

Simone Corbetta, Wim Meeus, Etienne Cappe, Francky Catthoor, and Agnes Fritsch

11.1 Motivation and Context

This chapter describes the proposed measurement-based modelling methodology to analyse the workload-dependent reliability of complex systems, and the implementation of this methodology in the HARPA context. The following relevant aspects will be addressed:

1. Workload dependency: The HARPA objectives are expressed in terms of well-known workloads specified by industrial partners and represented by their reference applications. Workload-dependency is required for accurate and realistic reliability model design [28, 38]. Nevertheless, for generalisation purposes, additional applications will be selected.
2. Temperature dependency: Harsh operating conditions will be taken into account since failure mechanisms (as NBTI) are exacerbated at high temperature (and voltage).
3. System-level: HARPA deals with parametric variation, but due to the sense/react nature of the project objectives, we will focus on a system-level perspective, in which the local effects (due to aging) are translated into functional failures at the processor boundaries, i.e. visible failures at the higher levels of the hardware/software stack.

S. Corbetta (✉) · W. Meeus · F. Catthoor
Imec, Leuven, Belgium
e-mail: simone.corbetta@polimi.it; wim.meeus@imec.be; francky.catthoor@imec.be

E. Cappe · A. Fritsch
Thales Commun. and Security (TCS), Gennevilliers, France
e-mail: etienne.cappe@thalesgroup.com; agnes.fritsch@thalesgroup.com

© Springer International Publishing AG, part of Springer Nature 2019
W. Fornaciari, D. Soudris (eds.), *Harnessing Performance Variability in Embedded and High-performance Many/Multi-core Platforms*,
https://doi.org/10.1007/978-3-319-91962-1_11

To be able to adapt the proposed methodology to the HARPA context, a strong experimental phase is required, analysing how the reference platform reacts to high-temperature operation. Notice that the choice of the reference hardware is driven by interest from industrial partners, while the need for a virtual platform overcomes the inadequate observability of internal nodes of the processor.

Finally, the methodology presented in this chapter addresses the scientific and technical outcomes of the HARPA project by means of

1. Definition of a general methodology to estimate the workload-dependent reliability from a black-box design, based on measurements in the field
2. Application of such methodology to the Freescale i.MX 6Quad SoC case [13] (one of our main reference platforms)
3. Characterisation of the impact of DVFS steps and policies on system-level reliability

The rest of this chapter is organised as follows. First, we will discuss related work in Sect. 11.2. Next, we will outline the different steps of our approach in Sect. 11.3. And then we will show the main results obtained in Sect. 11.4.

11.2 Related Work

11.2.1 Aging and NBTI

Reliability is a very important aspect of most practical systems [24]. An important category of hardware-induced mechanisms which strongly impact the reliability of modern microelectronic systems is aging of the nanoscale devices and interconnects [8, 9, 19]. Aging is a time-dependent process, after which characteristics of a device will deviate from nominal values [19]. We talk about parametric variation when referring to the (time-varying) variation of electrical characteristics of a device, such as gate delay or threshold voltage. This variation can then cause a functional failure in case the fault is not masked by the circuit. In scaled technologies, aging is getting worse [5] and experimental results have demonstrated that it is no longer feasible to constrain a circuit lifetime a priori, since device parameters are described by time-dependent distributions: Threshold voltage drifting ΔV_{th} is a stochastic process and, as such, it is not possible to determine a priori the lifetime of a device. Time-zero variability after the manufacturing process makes things even worse [33].

Different degradation mechanisms exist, such as negative-bias temperature instability (NBTI) [15, 18, 22], random telegraph noise (RTN) [25] and hot carrier injection (HCI) [30]. Their importance depends on the technology node (shrinking technology makes them worse) and operational parameters (temperature and supply voltage are acceleration factors).

NBTI in pMOS devices is known to be typically the dominant one [15, 18, 41]. The magnitude of degradation depends on a series of factors, both functional (frequency f and workload related) and non-functional (temperature T and supply voltage Vdd). Also analytical models and simulation approaches have been developed for NBTI [29]. It has been shown that atomistic models [15, 16, 18] not based on reaction-diffusion mechanisms are important to accurately capture the workload dependence [21]. Earlier models did not achieve this yet [2]. And the device level models have also been extended to larger datapath modules [20].

Next to devices also interconnects are prone to aging. In that case, electromigration is the main mechanism [6].

11.2.2 *Approaches to Deal with Aging and NBTI*

Several groups have recognised the need to deal with this aging not only at technology/device level but also at the architecture or system level, and this already for more than a decade [1]. Because of the strong temperature dependence of the aging mechanisms [6, 23, 39], thermal-aware design approaches abound [31, 34, 35].

But this is not sufficient on its own and also the workload-dependent effect should be exploited, as our own work has substantiated [27, 38]. This also involves so-called scenario-aware techniques [14, 32, 42]. These mitigation techniques are however not the focus of this chapter. They are reported in other parts of this book, especially Chaps. 4 and 6.

11.3 Measurement-Based Workload-Dependent Methodology

This section presents the methodology designed to fulfil the HARPA objectives related to the reliability model development. The HARPA methodology is based on strong experimental validation with the actual hardware platform (see Sect. 11.3.1) and the simulation framework (see later subsections). The need for a virtual platform accompanying the real hardware comes from the fact that the latter on its own doesn't allow observability of internal nodes of the processor, making it impractical to derive information with sufficient detail. Furthermore, the importance to maintain the real hardware as the reference architecture is dictated by the will to drive the research from real measurements and to meet the industrial interest in this direction. However, a mismatch in the simulated architecture and the real one is present (and known), and adds challenges to our research goal. Ultimately, the methodology presented in this chapter allows us to fulfil the following technical and scientific outcomes:

1. Definition of a general methodology to analyse the temperature- and workload-dependent reliability of a complex SoC, using a black-box approach, i.e. without requiring sensible and non-disclosed information from the SoC vendor
2. Definition of a general methodology to provide a system-level temperature- and workload-dependent reliability model (focusing on NBTI)
3. Application of the above methodology to a real commercial case, of industrial relevance: the Freescale i.MX 6Quad platform [13]

11.3.1 Freescale i.MX 6Quad SoC Platform

The Freescale i.MX 6Quad is a complex SoC that integrates many functionalities and IPs. A simplified block diagram taken from the reference manual [13] gives an idea of the complexity of the design, as shown in Fig. 11.1. The SoC hosts an ARM

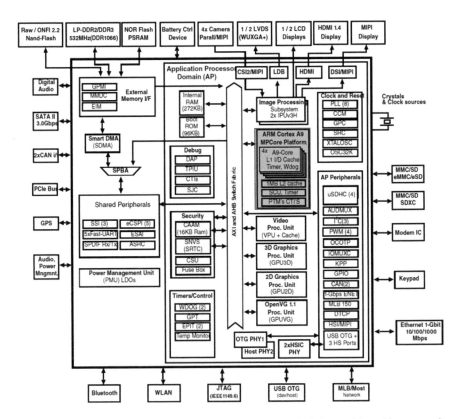

Fig. 11.1 Freescale i.MX 6Quad block diagram [13]. The ARM Cortex-A9 multiprocessor is highlighted in red

Cortex-A9 quad-core processor (the red-coloured box), implementing the ARMv7-A architecture. Data crunching applications can benefit from the on-chip presence of 3 GPUs and a video processing unit. Lots of peripherals support the flexibility of the device, and connect it to the external world.

Among others, the following SoC features are important to highlight [11, 12]:

- ARM A9-Core complex:
    - Symmetric CPU configuration.
    - Each core has a private 32kB L1 Icache and 32kB L1 Dcache.
    - Each core has a private NEON Media Processing Engine co-processor for SIMD, dual and single floating-point numbers arithmetic.
    - Unified 1MB L2 cache.

- Video Processing Unit and Image Processing Unit
- 2D/3D GPUs
- On-chip fast ROM and RAM memories
- Power management circuit
- Lots of peripherals

11.3.2 General Approach

The development of an accurate reliability model requires precise low-level information about the design, the architecture, microarchitecture and circuit of the target processor. Also, precise workload characteristics and waveforms must be available to provide an accurate workload profile. However, the Freescale chip design database is not available to customers and, even if we had the RTL design at hand, it would be simply impractical to simulate it in its entirety, due to the complexity of the design (in the order of a few million gates) and of the software application (execution time in the order of tens of minutes). For these reasons, the problem has been decomposed, and the proposed methodology to prevent this complexity is organised into three phases, as shown in Fig. 11.2.

11.3.3 Methodology Phases

The three phases are defined as follows:

1. Accurate modelling—this phase aims at determining precise timed workload from a selected subset of functional blocks. This is based on a simplified RTL design and an auxiliary simulation environments, as described in Sect. 11.3.3.1.
2. Approximation—this phase is used to generate precise gate-level waveforms starting from higher-level workload, as specified in Sect. 11.3.3.2.

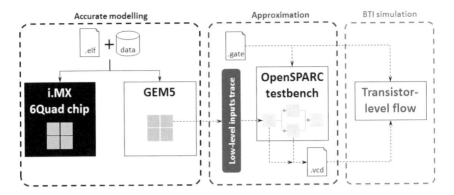

Fig. 11.2 Logical view of the HARPA estimation flow

3. BTI simulation—this phase computes the workload-dependent and temperature-dependent reliability metrics of the reference design, as specified in Sect. 11.3.3.3.

Each phase will be discussed in detail in the following sections.

11.3.3.1   Accurate Modelling

The purpose of this step is to generate workload (timed traces) at specified blocks of the processor, while running the reference application. Due to the complexity of the target processor and the application, timed traces are generated using a high-level model of the reference processor. This allows simulating the application and focusing on selected portions of the processor. We selected GEM5 as the architecture simulator [3] (freely available from www.gem5.org), for its ability to accurately model the ARMv7-A architecture and the ARM Cortex-A9 multi-processor. GEM5 is highly configurable, and comparison with the real hardware is performed to fine-tune the simulated platform configuration. Then we simulate the reference application, and keep track of the operands of predetermined instruction types (e.g. integer additions); they will represent the timed trace of data inputs to those functional blocks that support the execution of those instructions (e.g. the adder for the add/sub instructions). Figure 11.3 reports the flow to generate those traces.

The simulator has been modified to generate the timed list of operands and some additional access statistics about the execution time, the number of clock cycles, the IPC and the memory accesses for fine-tuning the simulation configuration. For the HARPA project objectives, the GEM5 simulator guarantees equivalence with the Freescale core, thanks to the presence of the ARMv7-A accurate model and the configurability of the simulator. The timing accuracy of the simulator also fits the project needs, since it remains below 7% absolute error [4].

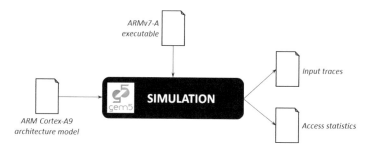

Fig. 11.3 GEM5 trace generation

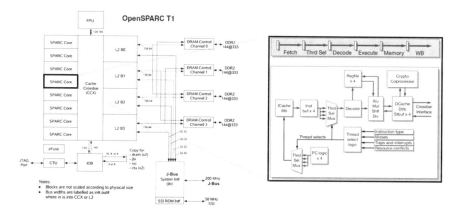

Fig. 11.4 OpenSPARC T1 block diagram and SPARC core microarchitecture

11.3.3.2 Approximation

A synthesizable RTL database is then required to generate low-level waveforms, synthesize the design and analyse the mapped netlist. However, very few freely available RTL designs exist, and the OpenSPARCT1 is the most complete one [37]. The OpenSPARCT1 is the open-source version of the Sun UltraSPARC processor, a complex SoC featuring 8 SPARCv9 cores, on-chip L2 banked cache and shared floating-point unit; the simplified block diagram is reported in Fig. 11.4, derived from [36]. The OpenSPARCT1 processor implements the SPARCv9 architecture. The block diagram of the processor and the microarchitecture of the cores is given in Fig. 11.4. It is composed of 8 cores with private L1 caches and a shared, banked L2 cache. It features classical pipelined cores with 6 stages [36]. The processor design and verification environment (RTL code, test-benches, scripts and ISA documentation) is freely available at http://www.oracle.com.

The SPARCv9 architecture [36] is different from the ARMv7-A architecture (http://infocenter.arm.com/help/index.jsp?topic=/com.arm.doc.ddi0406c/index.html). However, the arithmetic and logic operations make use of combinational or sequential circuits that we can assume being reasonably recurrent

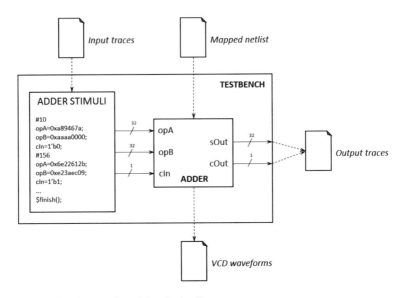

Fig. 11.5 Test bench example and timed stimuli

in the ASIC standard flow (e.g. think to the DesignWare library of components from Synopsys and how circuits are described in HDLs). This means that we can still use the RTL design of the SPARCv9 architecture, provided that we focus on isolated blocks of the data-path; this, in any case, is the portion of the processor where we expect to gain most from a reliability viewpoint, so the assumption fits very well with the objectives. Furthermore, this fits with the way we generate traces from high-level simulation (Sect. 11.3.3.1). Indeed, the HARPA methodology does not consider the entire core, but a number of key modules, focussing on the main floating-point and fixed-point arithmetic units. They have been chosen after engineering considerations about the design and analysing the final project objectives, reaching consensus with the industrial partners. Synthesis and simulation are performed on these modules in isolation, focusing on portions of the code that are more relevant to our purposes. With the OpenSPARCT1 design/verification environment available, it is now possible to simulate the desired application (represented by the collected traces) and generate low-level waveforms. An ad-hoc test-bench is created that instantiates the block under test and a vector-based stimuli specification. The inputs will be forced to known values, namely those from the collected traces. Control inputs are indirectly retrieved by analysing their role within the circuit. Then, the simulation will force the input values with appropriate timing information, and will let the internal nodes of the circuit settle to get node-level waveforms and output ports results. The overall picture is shown in Fig. 11.5.

11.3.3.3 BTI Simulation

The previous step aims at generating waveforms at internal nodes. These waveforms are now employed to gate-level BTI simulation and analysis. The collected traces are used as input to the transistor-level flow that computes workload-dependent reliability metrics of the circuit. Being dependent on the technology and the circuit, some calibration parameters will be required to drive the model development. The collection of points that will be later used for calibration is done through a first reliability experiments setup, discussed in more detail in Sect. 11.4.

11.3.4 Fault Detection and Analysis

Faults in a complex circuit can happen everywhere and they are potentially translated into failures, if they are not masked. The detection process is not trivial in the case of the HARPA project for a combination of reasons:

- Complexity is high and faults might be masked anywhere in the circuit: the observability is low in this case, and we might not even realize that a fault just occurred.
- The application deals with a large amount of data and faults are required to be detected within 1 ms from their occurrence, while the total execution time is in the order of minutes or even hours; this means that fault detection has to happen millions or billions of times during the execution of the application.

Thus, a lightweight fault checking mechanism is required to minimize the impact of error checking on the application behaviour: The more frequently we collect data the more the application will be influenced, but also the more likely a fault will be detected. In any case, faults can be detected by comparing the expected outputs to the real outputs of the system under test. This is accomplished using signatures to generate traces, a recurrent habit in the functional safety domain, e.g. in the automotive industry. This section describes how we detect faults that occur on the i.MX6 processor.

11.3.4.1 Offline Golden Trace Collection

The first step is then to generate (once) the trace of correct signatures. Different options exist to generate reference data: on the processor under test itself, on a different processor of the same type (e.g. on a second test board) or on a different processor (e.g. an x86_64). The first solution is not possible in practice because the cores on the device under test share the same DVFS settings which make them equally (un)reliable. Using a host computer is problematic too, as it turns out that minimal errors occur between the i.MX6 and the x86_64 architectures (e.g. floating-point rounding mismatch), while the outputs and the intermediate results need to be bit-wise identical for accurate comparison. Finally, generating the golden trace once offline on the same board, under nominal conditions (i.e. ambient temperature, default DVFS settings and default power supplier), is the most plausible solution. The golden trace is then stored for later comparison. Notice that the golden trace collection can be safely performed offline since it will not change with operational conditions (e.g. temperature or DVFS settings) that we may find in the run-time scenario.

11.3.4.2 Online Faults Identification

When the test and reference executions are in the same state, they can be compared for faults. Fault identification has to occur at run-time since we want to be able to detect faults within 1 ms from their occurrence. With the offline trace at our disposal, the setup in Fig. 11.6 is employed for online faults detection: The board will run a modified version of the code (to include signatures computation and sending), and the laptop will act as comparator.

The system under test and the laptop communicate through a TCP/IP network: The laptop waits for incoming data generated by the DUT board; signatures are buffered locally on the board and sent to the comparator by TCP/IP periodically. The reference application has been modified for this purpose:

- It computes signatures at specific configurable points (the period can be tuned such that it is in the $1\,\mu s$ to $1\,ms$ range) and sends these signatures to the comparator for storage (when recording a golden trace) or comparison (when running under stress).

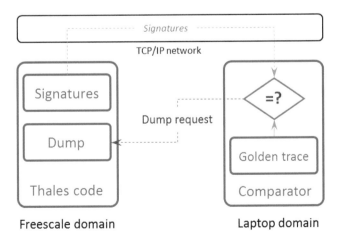

Fig. 11.6 Online fault detection setup

- It supports heap memory shadowing, i.e. memory allocation requests by the software (calloc()) are wrapped by code that uses ping-pong buffers, ensuring that internal application data are preserved for fault analysis if a fault is detected.
- It listens for a "mismatch detected" message from the comparator. If such message is received, the application is terminated and the internal data of the application is dumped to a file for further analysis.

As long as there is no mismatch, the application runs as before. As soon as the comparator detects a mismatch, a stop message is sent to the board. The application under test dumps its internal data to a file for the next debugging step, aimed at determining the causes of the failure.

11.3.4.3 Signature Accumulation

Signatures are 32-bit data computed through a simple accumulation algorithm that embeds the intermediate results computed by the processor. The simple accumulation algorithm performs the following steps:

1. Right rotate the current signature by 1 bit
2. Bitwise XOR the (rotated) signature with the new data

Signatures are initialised to a known value (e.g. 0xa5a5a5a5) and the accumulation procedure is applied periodically at specific points in the code.

11.4 Experimental Results

The purpose of the high-stress experiments is to measure the workload- and temperature-dependent reliability of the Freescale SoC, while running the desired application under harsh conditions. Data from these experiments will be used to calibrate the reliability model. Reliability in this context is measured by means of high-level metrics such as time-to-failure, where the first fault triggers the failure. Also, execution time variability plays an important role in real-time applications, and as such we keep track of the timing of the application under high-temperature conditions. Harsh conditions will both accelerate the faults process and increase the timing variability: The ambient temperature will be forced to a stable known value with the aid of a thermostream machine, while high voltage will be forced by modifying the DVFS kernel driver (this driver has been provided by Thales TCS).

Table 11.1 Selected modules from the OpenSPARC T1 design

Module	Number of equivalent gates
Integer ALU	6k
Integer multiplier	34k
Integer shifter	3k
Integer divider	13k
Floating-point ALU	29k
Floating-point multiplier	46k
Floating-point divider	11k

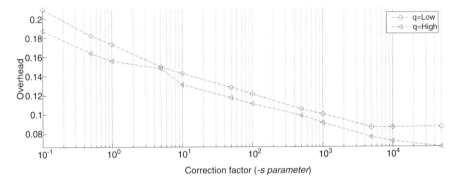

Fig. 11.7 Application overhead

11.4.1 Synthesis Results

The most important modules of the reference RTL design are present in Table 11.1. System clock frequency has been constrained to 900 MHz and only simple gates are employed (no 4-input gates, no special cells such as complex half- or full-adders) to fit with the BTI estimation back-end flow.

11.4.2 Signatures Accumulation Overhead

Signatures are first stored into a local buffer, which in turn waits for the maximum payload of a TCP packet to be reached. Signatures, however, are not sent to the buffer every time they are computed; they are instead sent once every number of iterations. This number is configurable, and the higher the number of signatures sent, the faster faults will be detected but the higher will be the overhead. Figure 11.7 reports the overhead as a function of the parameter that controls the signature sending period; this plot can be used to select the appropriate correction factor given the desired overhead maximum. The overhead is defined as the relative increase of execution time of the modified version from the original one.

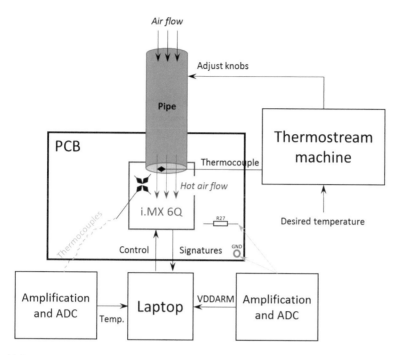

Fig. 11.8 Reliability setup

The overhead results are reported for both the low- and high-quality scenarios, for the Imec profile, considering performance DVFS governor. In general, the low-quality scenario accounts for a higher-performance overhead, namely 10% more on the average. This is due to the fact that the number of iterations per loop in the Low scenario is lower than in the High case, thus the total number of computed signatures is lower and the overhead is relatively higher. Also, the relative weight changes with DVFS settings. With the performance governor there is no actual dynamic adaptation, since the processor is then all the time kept at maximum frequency.

11.4.3 General Setup

The experiment setup is depicted in Fig. 11.8. Unless otherwise noted, this is the setup that is used during the following high-temperature experiments. The following components are visible:

- A Thermostream machine—to generate and blow hot air onto the chip's surface
- The i.MX 6Q chip mounted on the Freescale board [11]—running the reference workload

- The Amplification and ADC circuits—for thermocouples readings, and supply voltage trace analysis
- A Laptop—to collect raw data and control the board

11.4.3.1 Accurate Operating Temperature Control

The thermostream machine is pumped air operated, and the air temperature can be set using the knobs on the front panel. The desired temperature is stabilised by using a thermocouple at the end of the pipe, and the sensed value is sent back to the central unit for closed-loop adjustment. Thermocouples are also placed on the chip's surface at predefined positions to overcome the limitations of the digital sensor and to keep trace of the time-dependent evolution of temperature. Thermocouples are indeed able to identify gradients and trace temperature above 125 °C, while the thermal sensor can be used for absolute measurement below 125 degrees (i.e. the qualification and calibration point of the reference chip).

11.4.3.2 Voltage Stability Checks

Voltage supply to the ARM Cores is also traced. A voltmeter has been designed using the Arduino Uno platform, to trace VDD voltage. The VDD pin is taken from resistor R27 [10], while ground is taken from the board's ground pads. The analogue voltages are sent to the analogue input pins of the Arduino that is in charge of the digital conversion through a 10-bit ADC that guarantees 49 mV resolution. This allows to check whether the supply voltage is stable or not during the high-temperature experiments. This is of paramount importance to remove all the possible sources of noise that are not related to the NBTI phenomenon.

11.4.3.3 Sweeping Parameters

The detailed list of operating parameters that are set during the reliability experiment is as follows:

- Operating temperature: [125,170] °C. Motivation: The Freescale SoC is qualified for 125 degrees for the automotive industry [12]. Above this value, no guarantee remains that the chip will continue to work properly. The reference temperature is measured on the chip's surface.
- Supply voltage: [1.3,1.875] V in steps of 250 mV. Motivation: The Freescale SoC is qualified for 1.3 V [12]; 1.875 V is the maximum voltage we can have from the on-board power management IC, and 1.875 V is still in the NBTI range without causing hard breakdown.

- Running frequency: 996 MHz. Motivation: The maximum allowed by the platform, and the one for which the circuit has been designed for. This stresses the clock margins.

11.4.3.4 Temperature Sweep Procedure

During reliability experiments, the platform will be stressed at high temperature; the objective is to measure its reliability at multiple runs at steady-state temperature and without focusing on aging across multiple runs. The platform will then be stressed according to the following procedure:

1. At (relative) low temperature, power up the board and perform necessary platform initialisation
2. Power up the thermostream (refer to the setup from Fig. 11.9) and set the target temperature to the desired one
3. While performing the transition, run the synchronizer script. Notice that it will take a while to reach the steady-state point
4. Once the synchronizer script ends, run the application on the Freescale board and start collecting temperature values from the thermocouples.

The aforementioned procedure allows to focus on the steady-state operation point, and avoids running experiments during the temperature transient. The synchronizer is used to detect the steady-state temperature on the chip's surface; it is designed to end once this condition is reached. In Fig. 11.9, this is t_0 and it defines the reference time for fault time analysis. Indeed, at t_0 the application is finally started and t_1 determines the desired time to failure (if any). The steady-state detection is reached considering the fluctuation of the temperature values in a given window, considering a predetermined threshold value (in this case 0.250 degrees). Notice that a single iteration runs at most (i.e. in case of no faults) for a predetermined amount of time. Furthermore, the platform is reset between successive transitions, no matter whether a fault occurred or not.

11.4.4 Reference Application

For our experiments, we have used a representative reliability-sensitive application with real-time constraints from Thales TCS, namely the core of a spectrum sensing kernel [7] which is an important component of cognitive and opportunistic radio access technologies [40]. It sweeps the relevant frequency spectrum to obtain information about unused frequency bands. Based on this radio frequency, allocation can be performed. An embedded real-time application context has been developed for this experiment. The application relies heavily on computation-intensive signal processing in order to perform the sensing operations. It gathers samples from a large frequency range, performing filtering and splitting into frequency bands for

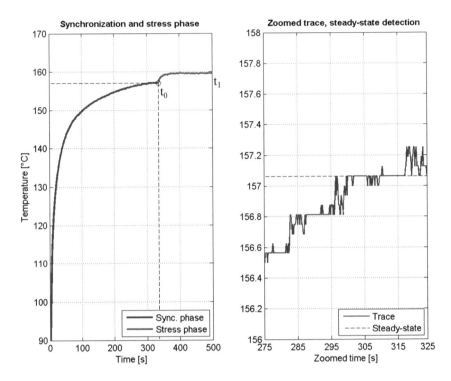

Fig. 11.9 Synchronizer example

further analysis. In our context, the sensing consists of three consecutive steps: wideband filtering for the signal operations, signal channelisation and parameter estimations. A 256-point Fast Fourier Transform (FFT) with data in complex long format is mainly used as the computation kernel.

11.4.5 Temperature-Dependent Execution Time

The NBTI-induced aging of devices makes the threshold voltage V_{th} increase temporarily whenever a trap captures a charge it becomes activated, and the net effect is a deterioration of the performance parameters of the PMOS transistor. When the trap emits its charge, the threshold voltage V_{th} returns (mostly) to its initial state. For instance, the gate delay increases as the temperature exceeds the specified operating boundaries. This is usually 125 degrees for automotive and industrial electronics. The increased gate delay may translate into either a fault or an increase in system delay. In the former case, if the delay exceeds the clock margins, the synchronous circuit can sample a wrong value and then eventually propagate it through the system. In the latter case, local delays might sum up and

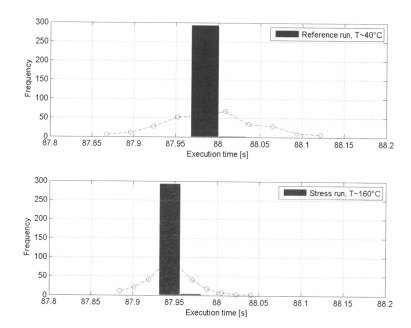

Fig. 11.10 Execution time comparison

lead to a system-wide increase in execution time. The purpose of these experiments is to measure this eventual delay at high temperature. We have run the reference application under constant temperature (set through the thermostream) and supply voltage (set through the DVFS governor) for 300 times, and collected the execution time of each iteration. For this experiment, we did not send signatures through the Ethernet connection, but instead stored the signature trace directly on the SDCard to avoid any unpredictable delay due to the TCP/IP stack. The comparison from Fig. 11.10 shows that there is no noticeable change in the distribution of the execution times at high temperature. Each plot reports the histogram (20 bins grouping) of 300 different measured samples, and the fitted distribution assuming population is Normal. The standard deviation is very small, and execution time samples are located within a ±100 ms range, i.e. ±0.1% with respect to the average. Nothing robust can be said about the slight shift of the high temperature average towards the left of the axis. Indeed, this is a negligible difference (less than 50 ms, 0.05% from the average).

In any case, another interesting result has been found comparing the signature traces of the previous runs. It has been found that a fault occurred in the processor, since a signature mismatch has been observed. Figure 11.11 reports the signed numerical difference between homonym signatures collected during the reference run (at 40 degrees) and during the stress run (at 160 degrees). It is clear that any positive or negative difference underlies a signature mismatch. The top-most part of the figure shows that there are mismatches in the first 60 iterations (~90 min of

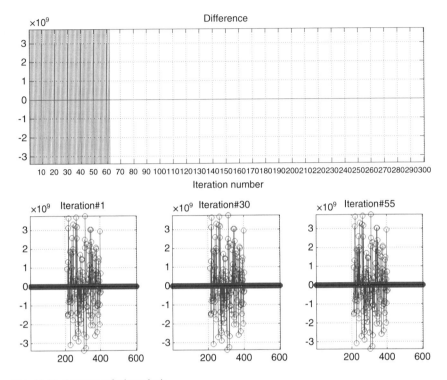

Fig. 11.11 Recurrent fault analysis

operation), and it shows also that the mismatch disappears at iteration #62. First of all, an in-depth analysis of the mismatches at different iterations has shown that these are recurrent at specific points in time, across successive iterations. Notice that each iteration re-reads the input data and re-computes the result. This means that the fault is localised at a precise point in the circuitry, and as soon as the application stresses that part, the computed result is impacted. However, it is not yet clear where exactly this fault has occurred, since the chosen signature sampling rate is too coarse-grained during this experiment. The very first fault occurred after 56 s of operation, so not immediately. This implies some "warming-up" period is present also.

Also, notice that the temperature is constant, the supply voltage did not experience flaws (see Fig. 11.12) and the frequency is constant too (execution time is constant). Thus, the conditions are good to match the results with the NTBI-induced aging assumption. Indeed, the only other changing parameter is temperature, and the fault expiration is perfectly reasonable in the context of the capture/emission theory.

Previous results suggest that a more detailed reliability campaign is worthwhile to be explored, as discussed in the next subsection.

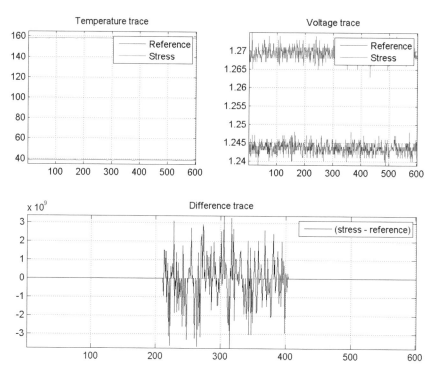

Fig. 11.12 Temperature and voltage traces

11.4.6   BTI Aging Experiments

NBTI is expected to play an important role at high temperatures. Early experiments (see Sect. 11.4.5) have shown that there are chances that a fault translates into functional failure. Further experiments are then required to analyze the system-level sensitivity to temperature of the entire Freescale chip, running the Thales spectrum sensing reference application (for workload-dependency). The purpose of this section is to provide initial results regarding the reliability of the Freescale chip as a function of increasing temperature, assuming the setup from Sect. 11.4.3. The following results will be shown:

1. System-level temperature dependency of failure time
2. Types of faults experienced in the Freescale platform
3. Analysis of the most sensitive (i.e. unreliable) portion of the reference application to high temperatures
4. Plans for next experiments

Note that due to the limited number of samples that can be obtained in a reasonable amount of time, any statistical projection of the reported data is actually not robust. In any case, the results show a clear trend, and can be used by industrial partners to focus on the most critical part of their application, for reliable products.

The time required to provide the BTI estimate depends on a number of factors, such as the number of transistors in the reference circuit, the topology of the circuit and the length of the waveform (in number of events).

11.4.6.1 System-Level Perspective

The setup presented in Sect. 11.3.4 allows to detect a system-level failure, for which an underlying failure mechanism holds. This section reports the measured time-to-failure at high temperatures on the Freescale device, while running the Thales TCS application. Figure 11.13 reports the measured time-to-failure against temperature.

On the left-hand side, one sample is reported from each run as a circle; crosses denote cases in which no fault has been detected. These are conventionally placed on the TTF axis at the maximum execution time, but they actually lie in the region above the striped line. We cannot safely assume their position, thus we do not consider them in our analysis. From this plot, it is clear that the probability of having faults increases with temperature (crosses overlap at relatively low temperature), and still dispersion is present, spanning almost two orders of magnitude. This result is expected, since it is known that NBTI-induced faults are widely spread along the time axis [17]. Still, there are cases in which the processor experienced no fault, despite the high temperature, to further demonstrate the variability of the underlying phenomenon. The right-hand side of Fig. 11.13 zooms in into the 168–172 degrees region, where the measured data has been grouped and analyzed. Box plots are used to determine the degree of dispersion of the data at different temperatures. The low samples count makes it harder to draw robust conclusions about any statistical

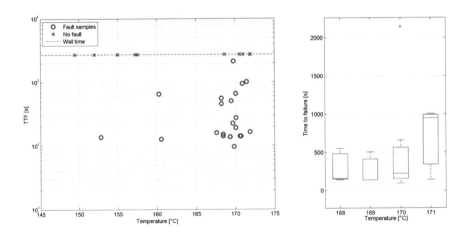

Fig. 11.13 Time-to-failure against temperature

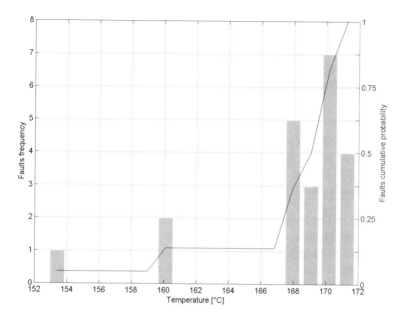

Fig. 11.14 Empirical cumulative distribution function and histogram

distribution, and also the range is not enough to make predictions. Furthermore, outliers are present, and this makes the fitting process even harder. In any case, it is evident that the temperature plays an important role in causing a fault in the processor. The empirical cumulative distribution function is given then in Fig. 11.14, along with the computed histogram. Notice that we experienced 21/34 (~62%) faulty cases against 12/32 (~38%) non-faulty cases as a summary.

11.4.6.2 Types of Faults

The adopted method guarantees the analysis of the failure time at system level, i.e. considering a functional fault as the point in which a fault manifests itself. Notice that the use of signatures (see Sect. 11.3.4.3) is capable of capturing the exact point in which a fault occurs, but this at the price of overhead. Instead, to minimize the impact on the workload, a smaller number of signatures is currently used, as also indicated in Sect. 11.3.4.3). Hence, now it is not possible to distinguish between the types of faults.

The following types of failures have been observed:

1. Functional failure—any mismatch in the signature that does not cause the platform to be reset (see next types) is considered a functional failure: somewhere in the circuit, the processor computed the wrong result.

2. Vdd failure—this group contains all those faults that cause the platform to be reset. The root cause of such faults is unknown since as soon as the platform is automatically reset, we lose the memory contents.
3. No failure—either no faults at all or masked faults fall in this group. We cannot distinguish them.

In the experiments reported here, most of the time the failure is actually a functional failure: In rare cases, we experienced a Vdd fault, and in only very few cases we were able to find the root cause. Indeed, it is known that the Freescale SABRE board has a voltage protection circuit that is placed close to the DC voltage plug [11]. The voltage protection circuit is composed of a diode that, when conducting, asserts a reset signal to the system and lights red; the circuit is designed so that the diode starts conducting when the input DC voltage is greater than 5.6 V [11]. However, we noticed that in some cases the diode starts conducting even when there is no major voltage peak, and it does so gradually (being a LED is easy to check whether it is conducting or not, and the intensity of the emitted light is an index of its bias). A thermocouple placed close to the diode has also shown that it starts conducting at high temperature, meaning that the reset signal is actually a spurious one, induced by the temperature increase through PCB and air from the main SoC. Note that the temperature at the diode is not found to be systematically high but depends on a number of factors including the direction of the pipe on the chip surface. For these types of faults, we do not provide quantitative values, since we are mostly interested in the functional ones in the HARPA context, but they are as important as the others from a system perspective.

11.4.6.3 Application Reliability Analysis

Another important analysis that can be conducted with the experimental setup proposed in this document relates to the most sensitive portion of the application to temperature, from a reliability standpoint. Indeed, it is expected that the application contains parts with different execution profiles, e.g. some of them may be integer intensive, while other floating-point intensive. This translates to a different impact on the system-level failures. The signatures contain also additional information of when in the code a fault occurred, by reporting an identifier representing the overall progress of the computation. This number is called the idBuffer, and will be used to perform post-process analysis for the back-end part of the flow (see Sect. 11.3.3.2). Figure 11.15 shows that there are actually portions that are more critical than the others, i.e. when a fault occurred they were more likely contributed to the signature mismatch.

Although each idBuffer covers a sequence of a number of instructions (dependent on the number of signatures selected) and not a single instruction, this is a useful tool to drive the post-processing phase. It also gives the partners an insight into where to focus their attention to increase the reliability of the final product.

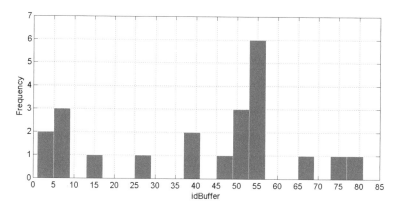

Fig. 11.15 idBuffer distribution during faulty runs

11.5   Conclusion and Future Work

This chapter has discussed the HARPA measurement-based modelling methodology for NBTI estimation, and has also introduced the way this has been adopted for the analysis of the reference platform. It has been shown that the accurate NTBI estimation requires calibration of the models against the reference design and the used technology. The experimental phase has been also successfully conducted, with the following noticeable outputs:

- Analysis of the temperature profile of the application/processor pair of choice
- Early analysis results of the reliability of the Freescale SoC as a function of temperature

The measured samples count is not enough to infer distributions in a statistically sound way, but they provide a deeper understanding of the processor and technology characteristics of the reference platform. They also provide a direction to follow for more elaborate future experiments.

References

1. Alam, M., Kang, K., Paul, B. C., & Roy, K. (2007). Reliability- and process-variation aware design of VLSI circuits. In *International Conference on Physics of Failure Analysis*.
2. Alam, M. A., & Mahapatra, S. (2004). A comprehensive model of PMOS NBTI degradation. *Microelectronics Reliability, 47*(6), 71–81.
3. Beckmann, B., Binkert, N., Saidi, A., Hestness, J., Black, G., Sewell, K., et al. (2011). The gem5 simulator. Tutorial at ISCA.
4. Beckmann, B., Binkert, N., Saidi, A., Hestness, J., Black, G., Sewell, K., et al. (2011). *The gem5 simulator.* Tutorial at ISCA 2011.

5. Borkar, S. (2005). Designing reliable systems from unreliable components: The challenges of transistor variability and degradation. In *IEEE Micro*.
6. Cheng, Y.-K., Tsai, C.-H., Teng, C.-C., & Kang, S.-M. S. (2002). Temperature-dependent Electromigration reliability. In *Electrothermal Analysis of VLSI Systems* (pp. 121–155).
7. Corbetta, S., Meeus, W., Rodopoulos, D., Cappe, E., Catthoor, F., & Fritsch, A. (2016, March). System-wide reliability analysis on real processor and application under Vdd and T stress. In *Proceedings of SELSE Conference, Austin*.
8. Deleonibus, S. (2009). *Electronic device architectures for the Nano-CMOS era*. Singapore: Pan Stanford Publishing.
9. Deleonibus, S. (2012). Prospects for nanoelectronics CMOS scaling and functional diversification. *ECS Transactions, 44*(1), 41–47.
10. Freescale Semiconductor. (2012). AN4509: i.MX 6Dual/6Quad Power Consumption Measurement.
11. Freescale Semiconductor. (2013). MCIMX6Q - Smart Device Board and Schematics.
12. Freescale Semiconductor. (2014). i.MX 6Dual/6Quad Automotive and Infotainment Applications Processors datasheet (IMX6DQAEC).
13. Freescale Semiconductors, Inc. (2014). i.MX 6Dual/6Quad Applications Processor Reference Manual (IMx6DQRM Rev 2).
14. Gheorghita, S. V., Palkovic, M., Hamers, J., Vandecappelle, A., Mamagkakis, S., Basten, T., et al. (2009). System-scenario-based design of dynamic embedded systems. *ACM Transactions on Design and Automation of Embedded Systems, 14*(1), 1–45.
15. Grasser, T., Kaczer, B., Goes, W., Reisinger, H., Aichinger, T., Hehenberger, P., et al. (2010). Recent advances in understanding the bias temperature instability. In *International Electronic Devices Meeting (IEDM) Conference*.
16. Grasser, T., Kaczer, B., Goes, W., Reisinger, H., Aichinger, T., Hehenberger, P., et al. (2011). The paradigm shift in understanding the bias temperature instability: From reaction/diffusion to switching oxide traps. *IEEE Transactions on Electron Devices, 58*(11), 3652–3266.
17. Grasser, T., Wagner, P. J., Reisinger, H., Aichinger, T., Pobegen, G., Nelhiebel, M., et al. (2011). Analytic modelling of the bias temperature instability using capture/emission time maps. In *International Electron Devices Modeling (IEDM) Conference*.
18. Kaczer, B., Grasser, T., Roussel, P. J., Franco, J., Degraeve, R., Ragnarsson, L. A., et al. (2010). Origin of NBTI variability in deeply scaled pFETs. In *IEEE International Reliability Physics Symposium (IRPS)*.
19. Kiamehr, S., Weckx, P., Tahoori, M., Kaczer, B., Kukner, H., Raghavan, P., et al. (2016, April). Impact of process variation and stochastic aging in nanoscale VLSI. In *Proceedings International Reliability Physics Symposium (IRPS), Pasadena*.
20. Kukner, H., Khatib, M., Morrison, S., Weckx, P., Raghavan, P., & Kaczer, B., et al. (2014). Degradation analysis of datapath logic subblocks under NBTI aging in FinFET technology. In *International Symposium on Quality Electronics Design (ISQED)*.
21. Kükner, H., Khan, S., Weckx, P., Raghavan, P., Hamdioui, S., Kaczer, B., et al. (2014). Comparison of reaction-diffusion and atomistic trap-based BTI models for logic gates. *IEEE Transactions on Device and Materials Reliability, 14*(1), 182–193.
22. Kumar, S. V., Kim, C. H., & Sapatnekar, S. S. (2006). An analytical model for negative bias temperature instability. In *IEEE ICCAD Conference*.
23. Luo, H., Wang, Y., He, K., Luo, R., Yang, H., & Xie, Y. (2007). Modeling of PMOS NBTI effect considering temperature variation. In *International Symposium on Quality Electronic Design (ISQED)*.
24. O'Connor, P., & Kleyner, A. (2012). *Practical reliability engineering* (5th ed.). New York: Wiley.
25. Realov, S., & Shepard, K. L. (2013) Analysis of random telegraph noise in 45-nm CMOS using on-chip characterization system. *IEEE Transactions on Electron Devices, 60*(5), 1716–1722.
26. Reisinger, H., Grasser, T., & Schlunder, W. G. C. (2010). The statistical analysis of individual defects constituting NBTI and its implications for modeling DC- and AC-stress. In *IEEE International Reliability Physics Symposium (IRPS)*.

27. Rodopoulos, D., Catthoor, F., & Soudris, D. (2015). Tackling performance variability due to RAS mechanisms with PID-controlled DVFS. *IEEE Computer Architecture Letters, 14*(2), 156–159.

28. Rodopoulos, D., Mahato, S. B., Camargo, V. V. A., Kaczer, B., Catthoor, F., Cosemans, S., et al. (2011). Time and workload dependent device variability in circuit simulations. In *International Conference on IC and Design Technology (ICICDT) Conference*.

29. Rodopoulos, D., Weckx, P., Noltsis, M., Catthoor, F., & Soudris, D. (2014). Atomistic pseudo-transient BTI simulation with inherent workload memory. *IEEE Transactions on Device and Materials Reliability, 14*(2), 704–714.

30. Rosenbaum, E., Rofan, R., & Hu, C. (1991). Effect of hot-carrier injection on n- and pMOSFET gate oxide integrity. *IEEE Electron Device Letters, 12*(11), 599–601.

31. Sankaranarayanan, K., Velusamy, S., Stan, M., & Skadron, K. (2005). A case for thermal-aware floorplanning at the microarchitectural level. *Journal on Instruction-Level Parallelism, 8*, 1–16.

32. Sanz, C., Prieto, M., Gomez, J. I., Papanikolau, A., Miranda, M., & Catthoor, F. (2008). Combining system scenarios and configurable memories to tolerate unpredictability. *ACM Transactions on Design Automation of Electronic Systems, 13*(3), Article Number 49.

33. Sarangi, S. R., Greskamp, B., Teodorescu, R., Nakano, J., Tiwari, A., & Torrellas, J. (2008). VARIUS: A model of process variation and resulting timing errors for microarchitects. *IEEE Transactions on Semiconductor Manufacturing, 21*(1), 3–13.

34. Skadron, K., Stan, M. R., Huang, W., Velusamy, S., Sankaranarayanan, K., & Tarjan, D. (2003). Temperature-aware microarchitecture. In *International Symposium on Computer Architecture (ISCA)*.

35. Skadron, K., Stan, M. R., Sankaranarayanan, K., Huang, W., Velusamy, S., & Tarjan, D. (2004). Temperature-aware microarchitecture: Modeling and implementation. *ACM Transactions on Architecture Code Optimization, 1*(1), 94–125.

36. Tremblay, M., & Chaudhry, S. (2008, February). A third-generation 65nm 16-Core 32-thread plus 32-Scout-Thread CMT SPARC(R) processor. OpenSPARC, Sun Microsystems.

37. Weaver, D. L., & Germond, T. (Eds.). (1994). *The SPARC architecture manual, Version 9*. Englewood Cliffs: SPARC International, Inc., Prentice Hall. ISBN 0-13-825001-4.

38. Weckz, P., Kaczer, B., Toledano-Loque, M., Grasser, T., Roussel, P. J., Kukner, H., et al. (2013). Defect-based methodology fork workload-dependent circuit lifetime projections - Application to SRAM. In *International Reliability Physics Symposium (IRPS)*.

39. Wu, E., Suñé, J., Lai, W., Nowak, E., McKenna, J., Vayshenker, A., et al. (2002). Interplay of voltage and temperature acceleration of oxide breakdown for ultra-thin gate oxides. *Solid-State Electronics, 46*(11), 1787–1798.

40. Yucek, T., & Arslan, H. (2009). A survey of spectrum sensing algorithms for cognitive radio applications. *IEEE Communications Surveys Tutorials, 11*(1), 116–130.

41. Zafar, S., Kim, Y. H., Narayanan, V., Cabral, C., Paruchuri, V., Doris, B., et al. (2006). A comparative study of NBTI and PBTI (charge trapping) in SiO_2/HfO_2 stacks with FUSI, TiN, Re Gates. In *Symposium on VLSI Technology*.

42. Zompakis, N., Bartzas, A., Soudris, D., & Catthoor, F. (2013). System scenarios-based architecture level exploration of SDR application using a network-on-chip simulation framework. *Microprocessors and Microsystems, 37*(6–7), 544–553.

Part VI
Applications

Chapter 12
Floreon+ Modules: A Real-World HARPA Application in the High-End HPC System Domain

Antoni Portero, Radim Vavrik, Martin Golasowski, Jiri Sevcik, Giuseppe Massari, Simone Libutti, William Fornaciari, Stepan Kuchar, and Vit Vondrak

12.1 Floreon+: HPC Domain Application

There are many types of natural disasters in the world, and many of them depend on the geographical location of the observed area. Floods are one of the worst and most recurrent types of natural disasters in Central and Eastern Europe. Floods [17] occur when discharges and water levels exceed their bank-full discharge values and overflow. Floods kill millions of people, more than any other natural disaster, and they are also the world's most expensive type of natural disaster.

Floods frequently affect the population and are therefore studied by numerous scientific research institutes. Almost all large rivers in Central and Eastern Europe have experienced catastrophic flood events, e.g., the 1993 and 1995 flooding of the river Rhine, the rivers Danube and Theiss in 1999 and 2002, the river Odra in 1997, the river Visla in 2001, and the river Labe in 2002. Floods, however, affect not only Central and Eastern Europe, but they represent a significant problem in many regions all around the world. The growing number of losses caused by floods in countries around the world suggests that global mitigation of disasters is not a simple matter, but rather a complex issue in which science and technology can play a significant role. Therefore, the issue of flood prediction and simulation has been selected as a case of choice for innovative development.

A. Portero (✉) · R. Vavrik · M. Golasowski · J. Sevcik · S. Kuchar · V. Vondrak
IT4Innovations, Ostrava - Poruba, The Czech Republic
e-mail: antonio.portero@vsb.cz; radim.vavrik@vsb.cz; martin.golasowski@vsb.cz; jiri.sevcik@vsb.cz; stepan.kuchar@vsb.cz; vit.vondrak@vsb.cz

G. Massari · S. Libutti · W. Fornaciari
Politecnico di Milano, Milano, Italy
e-mail: giuseppe.massari@polimi.it; simone.libutti@polimi.it; william.fornaciari@polimi.it

© Springer International Publishing AG, part of Springer Nature 2019
W. Fornaciari, D. Soudris (eds.), *Harnessing Performance Variability in Embedded and High-performance Many/Multi-core Platforms*,
https://doi.org/10.1007/978-3-319-91962-1_12

The project FLOREON$^+$ [15] started in 2006 as a research project funded by the regional government of the Moravian-Silesian region of the Czech Republic that required reliable models for flood simulations and predictions to minimise costs of post-flood repairs in areas impacted by severe floods. The principal goal of the research project FLOREON$^+$ (*FLOods REcognition On the Net*) is the development of a prototype modular system of environmental risks modelling and simulations. FLOREON$^+$ is based on modern Internet technologies with platform independence.

The FLOREON$^+$ project results help to simplify the process of disaster management and increase its operability and effectiveness. The main scopes of modelling and simulation are flood risk, traffic modelling during critical situations, water and air pollution risks, and other environmental hazards. Another efficient utilisation of the computing power could be computing What-If scenarios for decision support and to plan preventive actions. Part of the research employs the modelling of land cover, and land use changes based on thematic data collection (aerial photographs, satellite imagery), and application of the prediction tools brings attractive advantages to land use planning. Modelling the catchment response to severe flood events generates the opportunity to improve the set-up and dimensions of new channel systems, within the scope of hydrology and water management.

12.2 Experimental Application

The central thematic area of the project is hydrologic modelling and prediction. The system focuses on acquisition and analysis of relevant data in near-real time and uses this data to run hydrologic simulations with a short-term forecast. The results are then used for decision support in disaster management processes by providing predicted discharges on river gauges and prediction and visualisation of inundated areas in the landscape (Fig. 12.1).

In the simulation phase (Adaptivity in Simulations [2]) of the prediction cycle, adaptivity in the spatial resolution is essential to improve the accuracy of the result. Specifically, more computational resources are introduced when the weather looks more attractive (i.e. after hours or days of heavy rain). These may be used for computations that are triggered by stimulating activities detected in the forecast simulation. Or they may be part of the same simulation process execution if it has been re-engineered to use automatic adaptive accuracy refinement. In any case, the most accurate computation has to track the evolution of the predicted and actual weather in real time. The location and extent of finer results should evolve and move across the simulated landscape in the same way the actual weather is constantly moving.

The model used in this study is a rainfall–run-off model (RR) developed as part of the Floreon$^+$ project. Rainfall–run-off models [14] are dynamic mathematical models which transform rainfall to flow at the catchment outlet. The primary purpose of the model is to describe rainfall–run-off relations of a catchment area. Standard inputs of the model are precipitation measured by meteorological gauges,

Fig. 12.1 Inundation area for the Ostravice river

and spatial and physical properties of the river catchment area. Common outputs are surface run-off hydrographs which depict relations between discharge (Q) and time (t). Catchment areas for the model used for experiments in this chapter are parameterised by Manning's roughness coefficient (N), which approximates physical properties of the river basin, and CN curve values (CN), which approximate geological properties of the river catchment area.

The confidence intervals are constructed using the Monte Carlo (M-C) method where input data sets are sampled from probability distributions extracted from historical results and used as input for a large number of simulations. The modelling precision of the model uncertainty can be positively affected by increasing the number of M-C samples and can be determined by estimating the Nash–Sutcliffe model efficiency coefficient [18] between the original simulation output and one of the percentiles selected from the Monte Carlo results.

Only selected percentiles are extracted from all M-C simulation results and saved on the platform. These percentile simulations describe a possible development of the situation taking possible inaccuracies into account along with its probability. For example, 80th percentile specifies that there is an 80th probability that the real river discharge will be lower or equal to the simulated discharge based on historical data. These results can then be propagated further into the flood prediction process, for example, used as input for hydrodynamic modelling. Figure 12.2 shows the visualisation of simulated inundated areas based on RR uncertainty results.

The confidence intervals' accuracy can be positively affected by increasing the number of MC simulations (also referred to as *samples*), and can be determined by estimating the Nash–Sutcliffe model efficiency coefficient [18] between the original simulation output and one of the percentiles selected from the Monte Carlo results. The percentile simulations describe a possible development of the situation taking possible inaccuracies (along with their probability) into account.

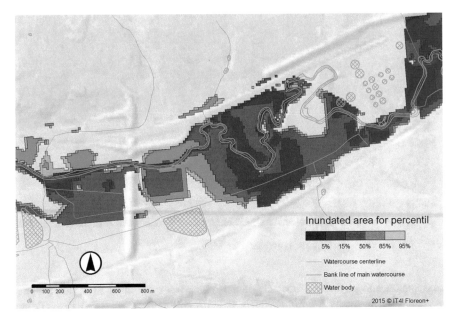

Fig. 12.2 Inundated area for percentile

12.2.1 Flood Warning Process

The flood warning process [24] can have two distinct stages: flood warning and response. The flood warning stage starts with a detection of the potential flooding threat. This activity should be done periodically based on river and precipitation monitoring and meteorology forecast. When a threat is detected, the flood warning committee is informed, and it issues a request for more detailed forecasts to the Institute of Hydrometeorology and local catchment area offices. These organisations provide

- Information about the actual river and reservoir situation
- Rainfall–run-off (RR) modelling simulation of surface run-off
- Hydrodynamic (HD) modelling flood lake simulations, flood maps, simulations of water elevation and water velocity, a real-time hydrological model for flood prediction using GIS, sediment transport, water quality analysis, etc.
- Erosion modelling simulation of water erosion
- Collection and archiving of flood data that can be used for estimating the magnitude of the flood based on historical evidence

If these predictions identify possible emergency situations, the flood warning committee alerts relevant agencies, and the process moves to the response stage. In this stage, countermeasures are implemented based on the forecast simulations from the flood warning stage. Also, flood predictions are still provided even in the response stage to support the decision processes related to performed countermeasures and actions. During the response stage, additional areas can be affected by the flood emergency, and these predictions should be able to identify these areas in advance.

We have integrated the most computationally demanding module of the Floreon⁺ platform with HARPA-OS to examine how HARPA-OS will influence its execution. The selected module provides uncertainty modelling of rainfall–run-off (RR) models. One of the inputs of such models is the precipitation forecast computed by numerical weather prediction models that can be affected by particular inaccuracies.

12.2.2   The Flood Forecasting Model

The RR models transform precipitation to water discharge levels by modelling individual parts of the rainfall–run-off process. Common inputs of these models are an approximation of physical properties of modelled river catchment and a time series of precipitations. Outputs of these models are represented by a time series of water discharge levels (i.e. the relation between water discharge (Q) and time (t)) for modelled parts of the river course. Our RR model uses the SCS-CN method [5] for transforming rainfall to run off with the main parameter curve value (CN) approximated from the hydrological soil group, land use, and hydrological conditions of the modelled catchments. The contribution from river segments to a sub-basin outlet is computed using the kinematic wave approximation parameterised by Manning's roughness coefficient (N), which approximates physical properties of the river channel using Manning's roughness coefficient (N).

$$E = 1 - \frac{\sum_{t=1}^{T}(Q_m^t - Q_s^t)^2}{\sum_{t=1}^{T}(Q_m^t - \overline{Q_s^t})^2} \tag{12.1}$$

Where:

Q_m is the measured discharge in a specific time-step.
Q_s is the simulated discharge in a specific time-step.
E is the Nash–Sutcliffe model efficiency coefficient.
$E = 1$ means that the simulation matches observed data perfectly.
$E = 0$ means that the simulation matches median of the observed data.
$E < 0$ means that the simulation is less precise than the median of the observed data.

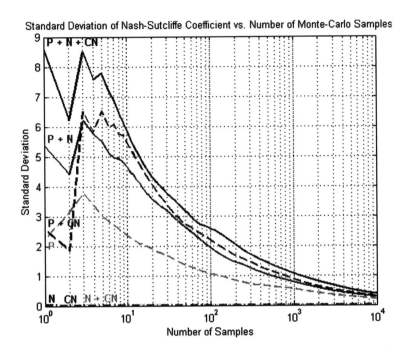

Fig. 12.3 Comparison of the precision provided by standard deviation of Nash–Sutcliffe coefficient and the number of Monte Carlo samples for different model parameters and their combinations

12.2.2.1 Uncertainties of the Rainfall–Run-off Model

The input precipitation data for short-term prediction are provided by numerical weather prediction models and can be affected by certain inaccuracies. Such inaccuracies can be projected into the output of the model by constructing confidence intervals. These intervals provide additional information about possible uncertainty of the model output and are constructed using the Monte Carlo (M-C) method. Data sets are sampled from the model input space and used as input for a large number of simulations. Precision of uncertainty simulations can be affected by changing the number of provided M-C samples. The precision of the simulations can be determined by computing the Nash–Sutcliffe (*NS*) model efficiency coefficient between the original simulation output and one of the percentiles selected from the Monte Carlo results. The *NS* coefficient is often used for estimating the precision of a given model by comparing its output with the observed data, but it can also be used for comparison between any two model outputs.

Figure 12.3 shows the precision of the simulated uncertainty results based on the number of Monte Carlo samples and different combinations of modelled parameters in the experimental model. *P* shows the standard deviation of uncertainty results where the only uncertainty of input precipitations was taken into account, *N* stands for the uncertainty model of the Manning's coefficient, *CN* for the *CN* uncertainty model, $N + CN$ describes the combination of Manning's coefficient and CN uncertainty models, and so on. These results show that the standard deviation is unstable for low numbers of M-C samples but starts to decrease steadily from around 5–8 samples. Precipitation uncertainty models and their combinations also show much higher standard deviations than CN and N based models. To obtain a sufficient precision for the simulation (i.e. minimise standard deviation of the result), the number of Monte Carlo samples for uncertainty simulation of the used Rainfall–Run-off model has to be in the order of 10,000 to 100,000. The sample count depends on the number of simulated model parameters and complexity of the model. Uncertainty simulation of CN and N could be executed for a much smaller sample count while maintaining sufficient levels of precision. This was mainly due to the lower sensitivity of the CN and N parameters when compared to the precipitation parameter, and also to the fact that precipitation uncertainty was sampled for each time-step of the simulation and each observed gauge, while CN and N did not depend on time.

12.2.2.2 On-Demand Simulations

Under the on-demand hydrologic simulations, a framework for running on-demand What-If Analysis (WIA) is created to simulate crisis situations. This also includes What-If hydrologic simulations. Through the web interface, users can create their own hydrologic What-If simulation running on this framework. They must choose the basic settings from the menu having the option to select river basin, schematisation, and rainfall–run-off model, for which the rainfall–run-off simulation and hydrodynamic model will be calculated. The framework allows the user to specify precipitation in selected precipitation stations and at selected times. This type of simulation also allows the user to edit the default parameters of sub-basins and channels. The next step is to run the simulation execution. Execution of What-If simulations is processed on an HPC cluster. Because the rainfall–run-off model is used as an input for the hydrodynamic model, the system allows the users to view hydrographs as soon as they are available, independently of the hydrodynamic computation. As soon as the hydrodynamic model computation is completed and the result values are stored within the spatial-temporal database, the users can view the simulated flood layer within the map interface together with the hydrographs.

12.2.3 Catchments Simulation

The proposed scenario monitors at runtime the behaviour of four concurrent instances of the uncertainty module. Each of these instances models the RR uncertainty for a different catchment of the Moravian-Silesian region: the Opava, Odra, Ostravice and Olza catchments (see Fig. 12.4). The watersheds are ordered

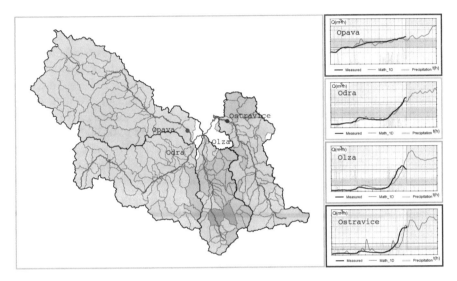

Fig. 12.4 The four main catchments (left) and outlet hydrographs (right) (the black line shows the measured discharge, the orange line shows the simulated discharge, X-axis: time in hours t(h), Y-axis: discharge, cubic meters per hour, Q(m^3/h))

according to the impact in case of flooding (the lower the index, the higher the importance):

- C_1: *Ostravice*—Functional urban areas with high population density and industrial areas in floodplain zones.
- C_2: *Olza*—Flood-sensitive zones in urban areas.
- C_3: *Odra*—Mountains in the upper part of the catchment can cause significant run-off. Less exposed urban areas.
- C_4: *Opava*—Soils with low infiltration capacity.

Each catchment is simulated independently, and individual instances do not interact with each other.

12.2.4 Application Scenarios

Based on the weather, FLOREON$^+$ can be subject to different service-level requirements. Indeed, the requirements can be translated to the parameters of the uncertainty modelling: A shorter response time in critical situations can, for example, be acquired by decreasing the number of MC samples—which, however, means reducing the precision of the results—or by allocating more computational resources to the application. Depending on the flood emergency situation, we identified three application scenarios that have different requirements. According to their criticality level, we tagged the scenarios as *standard*, *intermediate*, and *critical*.

12.2.4.1 Standard Operation

In this scenario, the weather is favourable, and the flood warning level is below the critical threshold. In this case, the computation can be relaxed, and some errors and deviations can be allowed. The system should only use as much power as needed for standard operation. Only one batch of Rainfall–Run-off simulations with uncertainty modelling has to be finished before the next batch starts. The results do not have to be available as soon as possible, so no excessive use of resources is needed. In this case, the estimated accuracy can be reduced.

12.2.4.2 Intermediate Operation

Due to the presence of limited precipitations, the forecast of discharge exceeds a warning threshold. In this case, in order to decrease the uncertainty of the model, the number of MC samples that must be performed by the simulation increases.

12.2.4.3 Critical Operation

Several days of continuous rain raise the water in rivers or a very heavy rainfall on a small area creates new free-flowing streams. These conditions are signalled by the river water level exceeding the flood emergency thresholds or precipitation amount exceeding the flash flood emergency thresholds. Much more accurate and frequent computations are needed in this scenario, and results should be provided as soon as possible. The number of MC samples increases, to decrease the uncertainty of the model.

12.3 HARPA-OS for HPC Environments

The HARPA-OS is a runtime resource manager (as presented in Chap. 4). Its role is to manage system resource allocation, taking into account both the status of the system resources and application requirements, combining pro-active and reactive strategies. For applications, performance requirements can vary not only among different applications but also during the execution of the same application. Aforementioned is a common scenario for HPC systems, where scientific applications make up most of the workload. It is a matter of fact that the performance requirements of this class of applications are often bound to the input data. This is due to the volume and type of data. For instance, monitoring systems acting at preventing natural disasters need to execute applications implementing mathe-matical models steadily, to analyse input data, detect possible criticality and notify the needs of operating accordingly. To guarantee to the applications the required-level performance, a conventional approach to HPC systems is to statically reserve computational resources, isolating the execution environment through virtualisation techniques. However, to reserve a fixed amount of resources (e.g. processors or entire nodes) can be ineffective, with the applications owning resources that are not fully exploited, all the time. Scaling the problem to the dimensions of an HPC centre, with several applications to host, we can face an overall under-utilisation of the system resources, leading to two issues: (1) the fragmentation of the available resources, and thus less space for further applications; (2) the loss of efficacy of power management techniques. The latter is due to the fact without a proper *consolidation* of the computational resources to allocate, we can have processors, or single cores, which are not fully exploited, but use enough time to avoid them to go into a deep-sleep state, wasting power saving opportunities.

12.3.1 The Runtime Resource Manager

In the context of this work, we employed the HARPA-OS runtime resource manager [12, 16] (Chap. 4). HARPA-OS enables the management of multiple applications that compete for the usage of multiple many-core computation devices [8]. It also proposes a runtime library [3, 21] that is in charge of (a) synchronising the execution of applications with runtime-variable resource allocations, and (b) notifying the resource manager of the runtime-variable Quality of Service goals (QoS) of applications. So that the HARPA-OS scheduling policy, which can be either chosen from a set of predefined ones or implemented from scratch, can take into account the feedback coming from applications when computing resource allocations.

We designed and implemented PerDeTemp (PERformance DEgradation TEM-Perature) Chap. 4, Sect. 4.6.2, a HARPA-OS scheduling policy that tries to meet the application performance requirements while minimising resource allocation [20]. When multiple computing resources are available, PerDeTemp employs a multi-objective heuristic to assign to applications only the healthiest and coolest cores. Such allocation aims at levelling the power flux over the whole chip, thus mitigating the ageing process and avoiding thermal hotspots [10].

Our framework is based on the idea of making the application terminate its execution just before the deadline (just-in-time, *jit*). This way, the amount of allocated processing elements is minimised. This, in turn, allows the resource manager to evenly level power consumption throughout the chip and to migrate the application to the coolest cores dynamically, thus evenly spreading heat and increasing the reliability of the silicon.

Figure 12.5 shows a comparison between the best-effort (maximises throughput and minimises execution time) of an application and our relaxed, deadline-aware

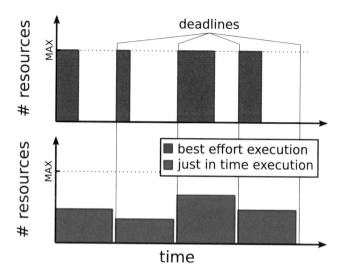

Fig. 12.5 Best-effort vs. *jit* scheduling

execution. Standard application execution is based on the idea of running the code as fast as possible (best-effort, *be*) using one thread per available processing element. This way, the results are available sooner; however, power consumption is maximised, and all the processing elements are stressed. Conversely, our just-in-time execution approach has works with the idea that, since applications can send feedback to the resource manager dynamically, resource allocation can be made more elastic: It is adjusted over time so that the runtime-variable performance demand of applications is always complied with, but the execution time of applications is still the maximum allowed one (i.e. applications terminate just before their deadline). The resource manager exploits the now-unused resources as a resource pool that can be used in multiple ways, e.g., to provide cool cores when the next resource allocation is computed.

The HARPA-OS runtime, which is linked, manages applications, and transparently monitors application execution statistics. Amongst these, one of the most important ones is the average throughput. It is worth noticing that each time HARPA-OS changes the resource allocation of an application, the runtime re-sets the throughput statistics; hence, the average throughput computed by the runtime always refers to the current resource allocation. It follows, then, that the average throughput is a very accurate predictor of how the application will behave (i.e. whether the application will terminate or not before the deadline) if the resource allocation remains constant until the application termination.

The current resource allocation provides the scheduling policy with feedback; applications use the HARPA-OS runtime library API to retrieve their current execution time and their average throughput. Basing on those values, the applications can compare their current performance, i.e. the average throughput under the current resource allocation, and the ideal throughput, i.e. the throughput that is needed by the application to terminate just before the deadline. This information is periodically sent back to the HARPA-OS as a performance gap:

$$gap_{performance} = \frac{throughput_{current} - throughput_{ideal}}{throughput_{ideal}} \qquad (12.2)$$

While **positive** *performance gaps* mean that the application is executing too fast, which in turn indicates that the HARPA-OS may seize some of the allocated processing elements and insert them into the pool of empty resources, in contrast, **negative** *performance gaps* indicate that the application needs more resources. In this case, the HARPA-OS takes some of the less stressed (i.e. healthier—cooler) processing elements (i.e. cores) from the unused resources pool and adds them to the set of resources that can be exploited by the application. Finally, *performance gaps* equal to 0 mean that the application is likely going to terminate just in time. Even in this case, however, the HARPA-OS may decide to change resource allocation, usually to swap the currently allocated set of processing elements with healthier and cooler ones.

12.3.2 HARPA Integration

To fully exploit HARPA-OS, we implemented the application in a way that allows its reconfiguration during runtime. From the HARPA-OS side, the features of the application are a set of resource requirements. Fig. 4.3, from Chap. 4, Sect. 4.3, where it is described summarizes the manageable execution model and shows the different methods that have to be supported by the application:

- *onSetup*: Setting up the application (initialize variables and structures, starting threads, performance goal, etc.).
- *onConfigure*: Configuration/re-configuration code (adjusting parameters, parallelism, resources, number of active threads, etc.).
- *onRun*: Single cycle of computation (e.g. computing a single rainfall–run-off simulation for one Monte Carlo sample).
- *onMonitor*: Performance and QoS monitoring. Check the current performance with respect to the goal.
- *onRelease*: Code termination and clean-up.

12.3.3 Hardware Infrastructure

After integrating the HARPA methods into our flood forecasting models, we deployed the enabled applications to a part of a supercomputer. The supercom-

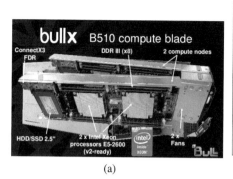 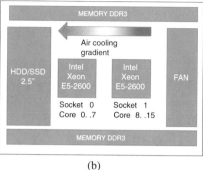

(a)   (b)

Fig. 12.6 The fan is on the right-hand side of the blade, creating an air cooling gradient between socket 0 and socket 1. (**a**) A photo of two Anselm Bullx B510 compute blades. (**b**) Schema of the blade

puter named Anselm (https://docs.it4i.cz/anselm-cluster-documentation/hardware-overview, June 2013) is an HPC cluster operated by IT4Innovations, the Czech national supercomputing centre, which contains 180 computational nodes without accelerators (i.e. GPU, Xeon Phi, FPGA). All these nodes have an interlink of high-speed fully non-blocking fat tree InfiniBand and Ethernet networks.

In our experiments, we used a chassis of 18 blades; the nodes are connected through InfiniBand. Each Anselm node is an Intel Corporation Xeon E5/Core i7. Each node of the system consists of 16 cores—2 Intel(R) Sandy Bridge E5-2665 @ 2.4 GHz CPU sockets each with eight cores and 20480 KB L2 cache. All nodes equip at least 64 GB DRAM.

The blades in the cluster are the B510 model; Fig. 12.6 shows a simplified representation of their air cooling system. Given that the fan is not equally distant from the two sockets, one socket is cooler than the other one. When the system is idle, the temperature difference between the two sockets is approximately 10 °C. The system has several monitoring tools installed such as power meters, *ganglia* [25] and *likwid* [26].

Figure 12.6 shows the temperature of one blade when it is idle; there is nothing more than the operating system running on it. This fact is important because, during the temperature experiments, we observed differences of 10° between the best and the worst cooled areas, the worst being furthest from the fan. For these experiments presented in Sect. 12.3.4, we defined three working temperature zones. When the temperature is low, it is represented on the heat map in blue. When the temperature is average, at the proper level, it is labelled with green. And when the temperature starts to be high it is represented with yellow, and then when quite high with red. All the temperatures are always under the limit that the vendor (Intel) recommends. Figure 12.7 presents the heat map of an Intel socket with 16 cores. The X-axis represents time and the Y-axis the cores. The temperature is in Celsius degrees. The *Diff. Per time step* is the range of temperature between the coldest and hottest core in a specific time. The *Median* is the average temperature of the cores of the socket

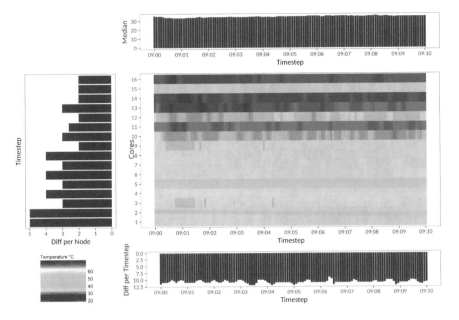

Fig. 12.7 Heat map of the two sockets of the blade in idle mode

in a specific time. The *Diff. per Node* is the maximum range of temperatures of a core during all the simulation.

12.3.4 Performance Analysis

The following experiments show the HPC solution to enforce runtime resource management decisions based on the standard control groups framework. A burst and a mixed workload analysis, performed on a multicore-based NUMA platform, reports some promising results both regarding performance, power saving, and ageing.

12.4 HARPA Testing in One Node

The following paragraph explains the execution of the models with HARPA-OS vs. non-HARPA-OS resource allocation. We run the models in the system with different governors. Governors that have a GNU/Linux Operating System (i.e. *performance* and *powersave*) by default set the frequency statically. The CPUfreq governor *powersafe* sets the lowest frequency from the scaling_min_freq border, and conversely the CPUfreq governor *"performance"* sets the highest frequency from the scaling_max_freq border. These two governors give the power/energy consumption range from a maximum of the system to minimum power consumption. Other GNU/Linux governors exist such as *userspace* and *ondemand* that provide intermediate solutions. These governors are developed to give the best-effort solution (minimum execution time) for the given range concerning power and frequency. But what would happen if our constraint is the execution time?

12.4.1 Execution of the Models with Constraints

To answer the above question, we performed two experiments. The first one involved execution of uncertainties of the rainfall–run-off model in the HPC system. The implementation simulated a real case of 12K MC samples in 10 min. The QoS with HARPA-OS must minimise the thermal footprint of the scheme. Reduction of the power consumption and execution of the model had to be under the time constraints. The non-QoS execution had no information about power, temperature or execution time. The implementation of the design was achieved in three ways, two non-QoS (a and b), and another with QoS constraints (c):

(a) Uncertainties of RR model in a non-managed HARPA-OS mode, only governed by the "performance" GNU/Linux governor. (b) Similar to (a) but with the "powersave" GNU/Linux governor. (c) Finally, uncertainties of RR model running in HARPA-OS [4] managed mode with the on-demand governor.

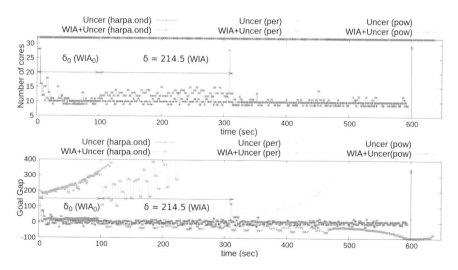

Fig. 12.8 Above: Number of resources allocated for performance, powersave and HARPA runtime with GNU/Linux ondemand governors during the execution of uncertainty models *(Uncer)* and Uncertainty and WIA concurrently *(Uncer+WIA)*. Below: Goal gap during execution time for Uncertainty model vs. Uncertainty more WIA using the three governors (per, pow, HARPA)

The second part of the experiment is similar to the previous one. However, the second application, the WIA, starts running a delta time equal to 100 s and evaluates 20 possible (What-If) positions. During this period, δ the WIA is requesting all cores of the system. So, it is producing resource allocation constraints to the execution of the uncertainty model. Figure 12.8 presents the number of cores allocated by the different governors. In the case of GNU/Linux governors, this value is always 32, the maximum number of virtual cores that the system has. In the case of Harpa governor, this value changes dynamically trying the goal gap to be zero. Figure 12.8 shows the δ time when WIA is also executed in the system. During this period, the resources (cores allocated) to perform the uncertainty model in the system have increased. Figure 12.8 shows the goal gap. There are three cases:

(a) *Performance*: The execution of the uncertainty model *Uncer(per)* ends long before the 10 min limit. Hence, the goal gap curve increases very quickly, even when the uncertainty and WIA run concurrently *WIA+Uncer(per)*. Then, during the δ time execution, there are some oscillations, which, however, go rapidly up since the model uses more resources than strictly necessary.

(b) *Powersave*: In this case, there are two situations which are different to when there is the execution of uncertainty model alone, and when the uncertainty and WIA operate concurrently. In the first instance, it is possible to observe that the goal gap is quite near-zero for the first period of the simulation (first 300 s). Later, step by step, the goal gap starts to increase, which means that again the resources are more than enough to execute the model in a shorter time than 10 min. However, in the second case *WIA+Uncer(pow)*, the goal gap became negative in the last fifty seconds of the simulation, which means that the resources given to this governor are not enough to execute both models in 10 min as requested. So, the execution missed the time deadline.

(c) *Harpa ondemand*: In this case, the runtime is minimising the goal gap as presented in section IV. The resources allocated (core #s are from 8 to 15, shown in Fig. 12.8) by the runtime are minimised as opposed to case (a) above, and the goal gap is oscillating around zero. In both cases, uncertainty (*Uncer (harpa.ond)*) and uncertainty more WIA (*Uncer+WIA (harpa.ond)*) are concurrently operating, and present similar behavior. The models finalise in the specified 10 min.

The curves presented in Fig. 12.8 show the different trade-offs of the monitors. In our case, the controls studied are power, energy consumption and temperature of the CPU sockets. Figure 12.9 presents the results related to the monitors. The monitoring of the system is performed adopting several GNU/Linux tools such as *ganglia*, the power meter *likwid*, and a per-core temperature driver with an installation of the GNU/Linux sensor library *lm-sensors*.

There is a trade-off concerning power consumption and heat. A middle QoS point exists between the non-QoS performance and power save points. Regarding relative values to QoS execution, the power consumption is reduced by 21% for one application and 24 % for two applications concerning the performance execution.

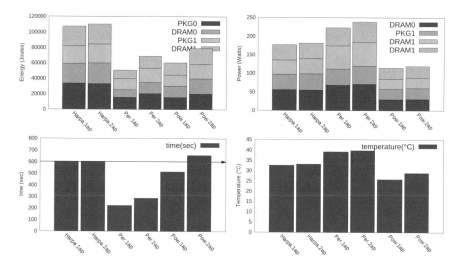

Fig. 12.9 Top-left: Total energy in joules for the six situations (3 governors with the two models (uncertainty model alone and uncertainty with WIA)), the energy can be divided in energy of the socket packages more the energy of the two modules DRAM. Top-right: Presents similar situations but power results in Watts. Bottom-left: Presents the execution time for the six situations, Bottom-right: The mean temperature of the 16th cores just before the end of the execution

Concerning temperature are 7 °C lower for both configurations than performance and 7 and 4 °C higher for one and two applications than power save execution. Because the execution time in performance execution takes 222 and 283 s, and there is no QoS monitoring, an overhead in resource utilisation is created. Therefore, the energy consumption of the system is lower than with QoS (HARPA-OS) execution.

12.4.2 Time Predictability with HARPA-OS

The main aim of the presented experiments is to show the reliability improvement in terms of time predictability [13]. The first experiment presented is only one example of possible execution of the model alone and with constraints (i.e. WIA model). But what can happen in terms of a reliable QoS for several running iterations, monitoring the power, energy and temperature of one system?

12.4.2.1 Power and Energy Monitors

Figure 12.10 presents the results of running the model one thousand iterations for 1 min and 1.2K MC samples. The figure presents results of running only the

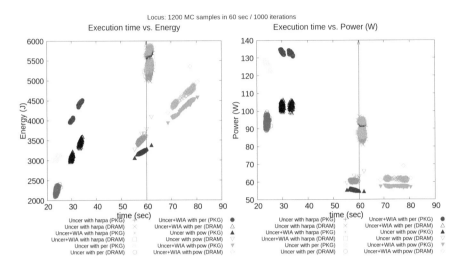

Fig. 12.10 Power and energy consumption vs. execution time in package (PKG) and memories (DRAM). Execution of 1K iterations of uncertainty model and uncertainty+WIA models. The iterations runs in three modes, two non-QoS named performance *per* and powersave *pow*, and one with QoS named *HARPA*

uncertainty model (*uncer*) and also running both models (*uncer+WIA*). The WIA starts execution of a random δ_0 from 0 to 15 s and performs two iterations (δ time).

Figure 12.10 shows the execution time vs. energy and power consumption in the chip package and DRAMs memories. The figure also reveals that all the iterations provide the results before or on the deadline of 1 min. Only the uncertainty with the WIA as a constraint in power save (*uncer+WIA with pow*) finalises after the term. The experiment reinforces the idea that there is a middle point regarding power consumption between the two GNU/Linux governors. The figure shows that with HARPA-OS activated the system has an overhead of 9–28% (J) with respect to the performance governor.

12.4.3 Execution of the Application Scenarios on the Platform

In this subsection, we run the uncertainty module on 16 nodes of an HPC cluster. The cluster presented has 18 blades. From all possible resources, we use 16 for our experiments, leaving two blades as a backup, log file system. The uncertainty module uses a hybrid OpenMP and MPI approach [22] to distribute the computations to multiple nodes.

Given that the performance requirements of the application are time-variable (e.g. low when sunny, intermediate/critical when rainy), the HPC centre may only allocate some computing nodes to the application and use the remaining ones

Table 12.1 For each experiment, number of Monte Carlo samples to be performed by the instance that models each catchment

	Thousand of MC samples to perform			
Experiment	C_1	C_2	C_3	C_4
α_1	1.5	1.5	1.5	1.5
α_2	1.5	3.5	5.0	7.0
α_3	7.0	7.0	7.0	7.0˙
β_1	3.5	3.5	3.5	3.5
β_2	7.0	7.0	7.0	7.0
β_3	3.5	7.0	12.0	15.0
γ	80.0 between all catchments			

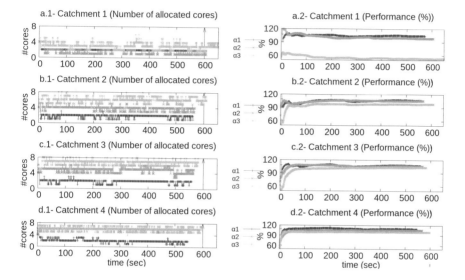

Fig. 12.11 Standard operation: light or no precipitation, low water level, (α experiments)

to execute other applications. Therefore, we performed our experiments in three different configurations: in the first one, we used only one node (i.e. standard operation). In the second, we used two nodes (i.e. intermediate operation). In the third, we used 16 blades of the cluster (i.e. critical operation).

Table 12.1 presents the set of experiments performed in the cluster. The experiments tagged α refer to the single-node scenario, while those tagged with β and γ respectively refer to the dual-node and 16 nodes are used from the entire cluster scenarios.

12.4.3.1   One-Node Configurations

Figures 12.11a.1, 12.11b.1, 12.11c.1 and 12.11d.1 show the number of cores allocated during the 10 min execution for each catchment, while Figs. 12.11a.2, 12.11b.2, 12.11c.2 and 12.11d.2 show the monitored performance

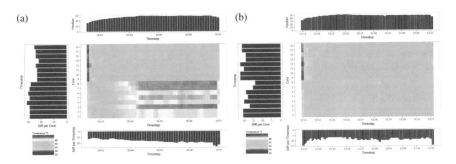

Fig. 12.12 Thermal map per core for a 10 min execution in one node. (**a**) Shows best-effort performance GNU/Linux governor (12K in total). (**b**) Shows the heat map in the case of HARPA PerDeTemp runtime with GNU/Linux performance governor (12K in total)

gap (g_p equations (12.2)) (as presented in Eq. (12.3), similar to (12.2), but in this case, the closer to 100% the gap is, the better the performance).

$$\text{gap}_{\text{performance}} = \frac{\text{throughput}_{\text{current}}}{\text{throughput}_{\text{ideal}}}(\%) \tag{12.3}$$

As shown by the experimental results, the number of allocated resources gets higher as the number of samples of MC to be performed increases. The allocation of resources is satisfactory for α_1 and α_2 scenarios, but not for α_3. In Fig. 12.11a.2, we can observe that the performance gap is below 100%, meaning that the resources required for this experiment are not enough. In this situation, a second node should be allocated.

Figure 12.12 presents the heat map of a single node when running an uncertainty module instance (12K MC samples) using a best-effort and just-in-time configuration, respectively. We obtained the heat maps by using the ganglia monitoring tool. In both figures, the X-axis represents time and the Y-axis represents the cores IDs.

- The *Diff per Core* (see the left part of the figures) is the difference in temperature per core c_i ($0 \leq i < 16$) during a period of time. In our case, the maximum difference of temperature of a core for a period of 10 min. Given a core c_i, and its maximum $t_{\text{max}}^{c_i}$ and minimum $t_{\text{min}}^{c_i}$ temparature, the *Diff per Core* $_i$ is $T_D^{c_i} = t_{\text{max}}^{c_i} - t_{\text{min}}^{c_i}$.
- The *Diff per Timestep* (the lower part of the figures) is the difference in temperature among all cores given a timestep. Given a time t and temperature T, the *Diff per Timestep* is $T_D^t = t_{\text{max}}^t - t_{\text{min}}^t$.
- The *Median* (the upper part of the figures) provides the median temperature per timestep. Given a time t, n equals the number of cores, and temperature T, the *Median* is $T_{\text{med}} = \Sigma_{c=0}^{c=n-1} = \frac{T_c^t}{n}$.

As can be seen in Fig. 12.12a (best-effort execution with CPUfreq performance governor) there is a hotspot in socket 0 [cores 0..7], which, as already shown in

Fig. 12.6, is the socket that is further from the fan. Conversely, Fig. 12.12b shows the execution with HARPA-OS-PerDeTemp. In this case, there are no hotspots and, thanks to the temperature-aware task migration performed by PerDeTemp, the temperature is perfectly distributed throughout all the available processing elements.

12.4.3.2 Two Nodes

Similarly, Fig. 12.13 shows the results for the dual-node configurations. In this case, β_2 has an equal number of samples to operate as α_3, and, since in this case, we have two computing nodes at our disposal, the computation is performed without issues. But the limit of 10 min is exceeded according to Fig. 12.13. Hence, in this case, the resources are again not enough for all the scenarios: Fig. 12.13b.2 shows that the performance gap of β_3 is below 100%.

12.4.3.3 Multi-Node Results

In this section, we show and discuss the results of the execution of the multi-nodes scenarios. On the application side, we used an MPI implementation of the Uncertainty model to enable the possibility of launching the application on multiple nodes. Moreover, for each node an instance of HARPA-OS was running, to manage the resource assignment locally. Again, we compared the HARPA-OS distributed execution of the workload against the unmanagedİ configuration with

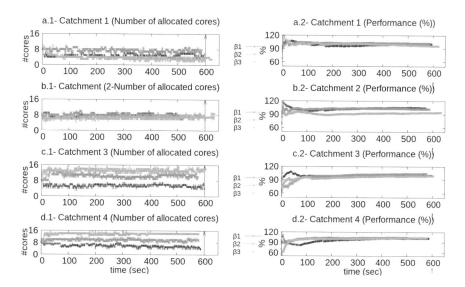

Fig. 12.13 Intermediate operation: intermediate execution, medium precipitation or water level warning threshold exceeded (β experiments) in two node execution

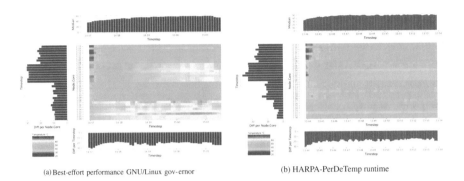

(a) Best-effort performance GNU/Linux gov-ernor   (b) HARPA-PerDeTemp runtime

Fig. 12.14 Thermal map per core for 10 min execution of the two node scenario. Case (**a**) shows unmanaged execution (performance CPUfreq governor). Case (**b**) Shows HARPA-OS (PerDeTemp) runtime management (performance CPUfreq governor)

the Linux Completely Fair Scheduler (CFS) scheduler in charge of managing the CPU resource allocation, and the CPUfreq governor statically set (performance or power save). As we have done for the previous scenarios, the comparison includes considerations regarding performance, temperature of the CPU cores, power end energy consumption. In the last experiment of this set, we executed the γ scenario on 18 nodes (execution is performed in nodes 1 to 16 with MPI $n = 16$, nodes 17 and 18 remain idle) with a requirement of 80K MC samples. This time, all the instances terminated just-in-time, i.e., exactly at the deadline (Figs. 12.14, 12.15, and 12.16). Using the *likwid* power monitor, we monitored the power consumption in both the *jit* and the *be* configurations. Whereas just-in-time execution leads to a maximum consumption of 100 W, the best-effort execution leads to a peak of 160 W per node (presented in Fig. 12.17). Therefore, we saved around 44% in maximum power consumption, while nevertheless complying with the deadline.

12.5 Evaluating HARPA Impact on Hardware Reliability

The Mean Time Between Failure (MTBF) is a well-known reliability metric, expressing the average time it takes for a failure to occur. It is calculated using the following formula: MTBF = (Total device hours)/(Total number of failures). The Failure Rate (FR) is the reciprocal of the MTBF:

$$FR = \frac{1}{MTBF} \qquad (12.4)$$

Acceleration Factors
The increase of the temperature of the test environment to which the devices are subjected can accelerate most failure mechanisms. It is, therefore, possible to

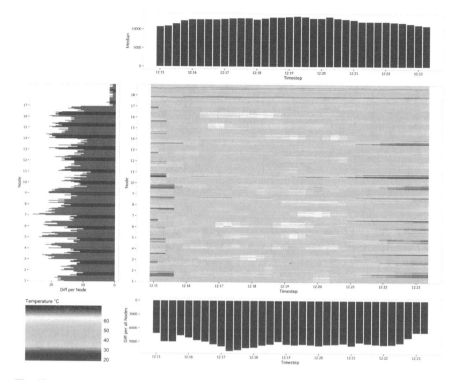

Fig. 12.15 HARPA-OS with PerDeTemp policy in 16 blades

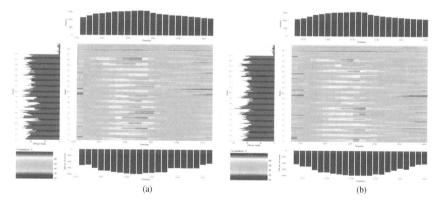

(a)   (b)

Fig. 12.16 Figures present hotspots (temperature peaks) even using the more conservative governor (power-save) in terms of power consumption. (**a**) Performance GNU/Linux governor in 16 blades. (**b**) Power-save GNU/Linux governor in 16 blades

perform relatively short tests which simulate many years of device stressing under more normal conditions. Obviously, it is necessary to have some measures of how greatly the failure mechanisms are accelerated. Suppose a device is stressed at a

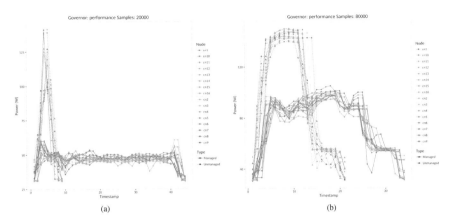

Fig. 12.17 Different behaviour in power consumption per blade. Performance governor presents peak consumption while PerDeTemp presents a decrease in peak performance. (**a**) Power consumption for 20K MC samples per blade. (**b**) Power consumption for 80K MC samples per blade

high temperature T_2, for time t_2. It is required to find the equivalent, t_1, which would cause the device the same level of stress at a lower temperature T_1. To this purpose, it is noted that if $T_2 > T_1$ then $t_2 > t_1$, we need to introduce an acceleration factor defined as follows:

$$A = \frac{t_1}{t_2} \tag{12.5}$$

hence, the ratio of the time at the lower temperature to that at the higher. Since the acceleration factor is a dimensionless ratio, it will be equal to the ratio of any other parameters proportional to the stress times. The most useful ratio is that of failure rates, which are inversely proportional to stress times if FR_1 is the failure rate at T_1, and FR_2 the failure at T_2 then:

$$T_2 > T_1; FR_2 > FR_1; A = \frac{t_1}{t_2} \tag{12.6}$$

So if FR_2 is known, FR_1 is found simply as $FR_1 = A \times FR_2$ All that is now required is to find the value of the acceleration factor.

Temperature Acceleration

The High-Temperature-Bias (HTRB) test is an example of a temperature-accelerated test. It is usually found that the rates of the reactions causing device failure are accelerated with temperature according to the Arrhenius equation:

$$\text{Acceleration}_{\text{factor}} = e^{\frac{E_t}{k(\frac{1}{T_1} - \frac{1}{T_2})}} \tag{12.7}$$

where E_A = activation energy of the failure mechanism, k (Boltzmann constant = 8.6E−5 eV/K), A, T_1, T_2 are as defined above. The activation energy E_A is found experimentally, and is usually in the order of 1.0 eV, depending on the predominant failure mechanism. For the failure mechanisms for small silicon technologies, a value of 0.3 eV has been determined.

Mean Time Between Failures (MTBF)

The example is given for two sockets of 8 cores each in ANSELM. According to Sect. 12.3.2, both sockets have a difference of several °C when the chip is idle. We estimated the MTBF for 14K MC samples goal per 10 min. This situation can be the standard usage of the system. From the set of temperatures, we take the per core average minimum temperature T_1 and maximum temperature T_2 for the two execution models, the HARPA PerDeTemp policy and the best-effort performance GNU/Linux. The difference between socket 0 and socket 1 is illustrated in Sect. 12.3.3. The position of the fan on one side of the blade produces a gradient in temperature of around 10 (°C) degrees between sockets (Table 12.2).

$$Relative_{acceleration\ factor} = \frac{Acceleration_{factor(best\text{-}effort)}}{Acceleration_{factor(PerDeTemp)}} \tag{12.8}$$

According to Eq. (12.7), we have used a range of E_A values from 0.3 to 0.9 eV, the values which are taken as lower and upper limits, and the real value is in the range. Therefore, we computed the AF and the MTBF for the two sockets 0, 1 and the two policies. The results are exhibited in Table 12.3.

Table 12.2 Temperatures in Kelvin ($T_{(K)} = T_{(°C)} + 273.15$)

	Socket 0	Socket 1
Best-effort		
T_{max}	340.525	324.525
T_{min}	328.4	313.025
PerDeTemp		
T_{max}	327.525	318.65
T_{min}	321.775	310.775

Table 12.3 Acceleration factor (AF) and MTBF for two different cooling systems and two possible E_A

	0.3 eV	0.9 eV
AF best-effort, socket 0	1.46	3.11
AF best-effort, socket 1	1.48	3.27
AF PerDeTemp, socket 0	1.21	1.77
AF PerDeTemp, socket 1	1.32	2.30
MTBF best-effort, socket 0	0.69	0.32
MTBF best-effort, socket 1	0.67	0.31
MTBF PerDeTemp, socket 0	0.83	0.56
MTBF PerDeTemp, socket 1	0.76	0.44

Table 12.4 Relative MTBF best-effort vs. PerDeTemp

MTBF best-effort/PerDeTemp, socket 0	0.83	0.57
MTBF best-effort/PerDeTemp, socket 1	0.89	0.70

Reliability Trade-Offs

The presence of hotspots is known to have adverse effects on the Mean Time Between Failures of the system (MTBF) [19], and can be mitigated [23] through the migration of tasks from one processor to a different in a Chip MultiProcessor (CMP) system, if the processor heats up to a threshold value. The ageing acceleration factor depends on the difference (Δ) of temperature.

According to the MTBF (Table 12.4) estimation presented in [9], running the application in a HARPA-OS runtime with a PerDeTemp configuration improves the reliability of the system from 17% ($1 - \text{MTBF}_{\frac{\text{best-effort}}{\text{PerDeTemp}}} = 1 - 0.83, eV = 0.3$), Table 12.4) to 43% (($1 - \text{MTBF}_{\frac{\text{best-effort}}{\text{PerDeTemp}}} = 1 - 0.57, eV = 0.9$)) in the case of poor cooling (socket 0). With a better cooled (socket 1), the MTBF for the best-effort relatively grows, but it is still 11% ($1 - \text{MTBF}_{\frac{\text{best-effort}}{\text{PerDeTemp}}} = 1 - 0.89, eV = 0.3$) to 30% ($1 - \text{MTBF}_{\frac{\text{best-effort}}{\text{PerDeTemp}}} = 1 - 0.7, eV = 0.9$) better if we manage the system with PerDeTemp (*jit*) instead of using the best-effort (*be*) policy.

12.5.1 Discussion of Applicability and Quality of the Results

The model used for the project is based on the Monte Carlo simulation. The Monte Carlo simulation is extensively used by the HPC community because it can be employed to generate massively parallel programs [6, 7]. The Monte Carlo simulation and its uncertainty are de facto fault tolerant since the algorithm can provide an answer even if some samples are lost. The rainfall–run-off model, its calibration [1, 27], and uncertainty must be executed every period that new data arrives from weather stations to provide as accurate results as possible. Hence, the QoS/time constraints allow relaxing of the execution requirements.

Since critical (alarm) situations are sporadic, instead of executing the model in best-effort, minimising the time but maximising the power consumption, running it with the QoS/time constraints is possible, as the results of the models must be provided before the deadline, otherwise it is too late. Our HPC models fall into the taxonomy of algorithms for urgent computing [28]. Therefore, the methodology can be reused for any program that falls under this classification. Thus, when the constraints are not very tight, it is possible to run the models allocating minimum resources and in the less degraded (healthier) parts of the system. HPC systems for urgent computing need very high availability and reliability. We have shown in this document the trade-offs between running the algorithm on the system in best-effort versus the Harpa-OS runtime. The balanced allocation of threads in the cores is taking into account which cores are more degraded, improving the reliability and hence expanding the lifetime of the system.

12.5.2 Conclusions

The chapter describes extensive testing and evaluation of representative applications of the HPC domain running with HARPA runtime support. The evaluation of the case studies has demonstrated the suitability of the techniques and methods developed in the HARPA project to both targeted compute domains (embedded and HPC). Benchmarking activities are performed as well, with reference platforms used to host these applications without HARPA framework. Validation of the HARPA environment verifies the fulfilment of our objectives concerning the HARPA environments functionality by monitoring and evaluating the performance, scalability and availability of adapted applications with a focus on efficiency in providing specified QoS levels in the presence of severe process variations. QoS levels are evaluated before and after the integration to enable the exact comparison of provided benefits. During the HARPA project, we have tried to assess the validation of the HARPA environment on HPC systems.

We have illustrated the advantages of running the HPC applications regarding availability and reliability. We have presented experiments that demonstrate the efficiency of providing specified QoS level even in the presence of degradation in the system (i.e. bubble). Yet still more work can be done concerning energy benefits. The current HPC systems are memory dominated. The power measurements are performed at the level of the socket, and inside the socket with the CPU, there are different layers of memory (L3, L2, L1 and LLC). According to our measurements, decreasing the Integrated Circuit (IC) power-supply pin V_{dd} with Dynamic Voltage and Frequency Scaling (DVFS) provides benefits regarding lowering the power, but not in reducing the global energy consumed during execution. This could be because the leakage power (static power) is dominant or has a high percentage of the consumption of the global system. Current monitors cannot provide us information about power consumption in the cores, only at the level of the socket. So, we think more experimentation is needed to ascertain what percentage of power consumption is due to the memory layer (mainly main memory) versus the CPUs in the socket. These tests could help us to disable or decrease the power consumption in memory parts that are underused, and in this way observe real energy savings.

In this chapter, we presented a framework that, based on the runtime-variable computational resources demand of applications, strives to minimize the resource usage of applications while making them comply with their deadlines. We call this practice *just-in-time execution* since the termination of the application is always as near as possible to the deadline. The framework can use the pool of unused [11] resources in multiple fashions, e.g., to swap allocated faulty processing elements with healthier ones, or to periodically assign to applications cool processing elements, hence evenly levelling the generated heat throughout the chip even in cases of asymmetrical cooling.

Acknowledgements This work was supported by The Ministry of Education, Youth and Sports from the National Programme of Sustainability (NPU II) project "IT4Innovations excellence in science—LQ1602" and by the IT4Innovations infrastructure which is supported from the Large

292 A. Portero et al.

Infrastructures for Research, Experimental Development and Innovations project "IT4Innovations National Supercomputing Center LM2015070". The work was also supported by the European Union FP-7 program through the HARPA project (grant no. 612069).

References

1. Abulohom, M. S., Shah, M. S., & Ghumman, A. R. (2001). Development of a rainfall-runoff model, its calibration and validation. *Water Resources Management, 15*(3), 149163.
2. Bahremand, A., & De Smedt, F. (2008). Distributed hydrological modeling and sensitivity analysis in Torysa watershed, Slovakia. *Water Resources Management, 22*(3), 293408.
3. Bellas, P., Massari, G., & Fornaciari, W. (2012). A RTMR proposal for multi/many-core platforms and reconfigurable applications. In *ReCoSoC*.
4. Bellasi P., Massari, G., & Fornaciari, W. (2015). Effective runtime resource management using linux control groups with the barbequertrm framework. *ACM Transactions on Embedded Computing Systems, 14*(2), 39:139:17.
5. Beven, K. (2012). *Rainfall-runoff modelling*. New York: John Wiley & Sons.
6. Billinton, R., Allan, R.N., IEEE Power Engineering Society. Power Engineering Education Committee, & IEEE Power Engineering Society. Power system Engineering Committee (1989). Reliability assessment of composite generation and transmission systems. IEEE Power Engineering Society Tutorial, 90EH0311-1-PWR.
7. Billinton, R., & Li, W. (1994). *Reliability assessment of electric power systems using Monte Carlo methods*. New York: Plenum Press.
8. Bolchini, C., Carminati, M., Gribaudo, M., & Miele, A. (2014). A lightweight and open-source framework for the lifetime estimation of multicore systems. In *IEEE 32nd International Conference on Computer Design (ICCD)*.
9. Ellerman, P. (2012). Calculating reliability using FIT & MTTF: Arrhenius HTOL model. In MICROSEMI, Technical Report.
10. Ganeshpure, K., & Kundu, S. (2014). Performance-driven dynamic thermal management of mpsoc based on task rescheduling. *ACM Transactions on Design Automation of Electronic Systems, 19*(2), 11:1–11:33.
11. Hardy, D., Sideris, I., Ladas, N., & Sazeides, Y. (2012). The performance vulnerability of architectural and non-architectural arrays to permanent faults. In *MICRO* (pp. 48–59).
12. Harpa harnessing performance variability fp7 project (2013). http://www.harpa-project.eu
13. Hartman, A. S., & Thomas, D. E. (2012). Lifetime improvement through runtime wear-based task mapping. In *Proceedings of the Eighth IEEE/ACM/IFIP International Conference on Hardware/Software Codesign and System Synthesis, CODES+ISSS'12* (pp. 13–22). New York: ACM.
14. Golasowski, M., Litschmannova, M., Kuchar, M., Podhoranyi, M., & Martinovic, J. (2015). Uncertainty modelling in rainfall-runoff simulations based on parallel Monte Carlo method. *International Journal on Non-standard Computing and Artificial Intelligence NNW, 25*(3), 267–286.
15. Martinovic, J., Kuchar, S., Vondrak, I., Vondrak, V., Sir, B., & Unucka, J. (2010). Multiple scenarios computing in the flood prediction system FLOREON. In *European Conference on Modelling and Simulation* (pp. 182–188).
16. Massari, G., Libutti, S., Portero, A., Vavrik, R., Vondrak, V., & Fornaciari, W. (2015). Harnessing performance variability: A HPC-oriented application scenario. In *Euromicro Conference on Digital System Design (DSD)*, Funchal, Madeira, Portugal.
17. Meesuka, V., Vojinovica, Z., Mynetta, A. E., & Abdullahd, A. F. (2015). Urban flood modelling combining top-view LiDAR data with ground-view SfM observations. *Advances in Water Resources, 75*, 105–117.

18. Nash, J., & Sutcliffe, J. (1970). River flow forecasting through conceptual models part I. A discussion of principles. *Journal of Hydrology, 10*(3), 282290. https://www.doi.org/10.1016/0022-1694(70)90255-6. Procedia Computer Science

19. National Semiconductor. (2000). Understanding integrated circuit package power capabilities. Available: www.national.com

20. Oh, D., Kim, N. S., Chen, C. C. P., Davoodi, A., & Hu, Y. H. (2010). Runtime temperature-based power estimation for optimizing throughput of thermal-constrained multi-core processors. In *Proceedings of the 2010 Asia and South Pacific Design Automation Conference, ASPDAC'10*, Piscataway (pp. 593–599). New York: IEEE Press.

21. Portero, A., Kuchar, S., Vavrik, R., Golasowski, M., & Vondrak, V. (2014). System and application scenarios for disaster management processes, the rainfall-runoff model case study. In *CISIM* (pp. 315–326).

22. Portero, A., Sevcík, J., Golasowski, M., Vavrík, R., Libutti, S., Massari, G., et al. (2016). Using an adaptive and time predictable runtime system for power-aware HPC-oriented applications. In *IGSC 2016* (pp. 1–6).

23. Powell, M. D., Gomaa, M., & Vijaykumar, T. N. (2004). Heat-and-run: leveraging SMT and CMP to manage power density through the operating system. In *ASPLOS*.

24. Qiua, L., Dua, Z., Zhua, Q., & Fane, Y. (2017). An integrated flood management system based on linking environmental models and disaster-related data. *Environmental Modelling & Software, 91*, 111–126.

25. Sacerdoti, F. D., Katz, M. J., & Massie, M. L. & Culler, D. E. (2003). Wide area cluster monitoring with Ganglia. In *CLUSTER* (Vol. 3, pp. 289–289).

26. Treibig, J., Hager, G., & Wellein, G. (2010). Likwid: A lightweight performance-oriented tool suite for x86 multicore environments. In *Proceedings of the 39th International Conference on Parallel Processing Workshops (ICPPW)* (pp. 207–216). New York: IEEE.

27. Vavrik, R., Theuer, M., Golasowski, M., Kuchar, S., Podhoranyi, M., & Vondrak, V. (2015). Automatic calibration of rainfall-runoff models and its parallelization strategies. *AIP Conference Proceedings, 1648*, 830014. https://aip.scitation.org/doi/abs/10.1063/1.4913040

28. Yoshimoto, K. K., Choi, D. J., Moore, R. L., Majumdar, A., & Hocks, E. (2012). Implementations of urgent computing on production HPC systems. *Procedia Computer Science, 9*, 1687–1693.

Chapter 13
Beesper SmartBridge: A Real-World HARPA Application in the Low-End Embedded System Domain

Federico Sassi, Alessandro Bacchini, and Giuseppe Massari

13.1 Beesper SmartBridge

Beesper is a flexible system for wireless distributed data collection, which allows to measure, visualize and analyse environmental parameters of various types, including: ambient temperature, pressure, humidity, light intensity, noise, air quality, vibrations and crack size, amongst others. In particular, a Beesper system is composed by a network of low-power wireless sensor nodes and one, or more, bridge node that collects the data and sends it to a remote server for further analysis. In order to be deployable in any environment, all components of a Beesper system are battery operated and guarantee multi-year monitoring activity. This requirement limits the computational power of the Beesper Bridge and the type of analysis on the collected data. In order to overcome this limitation, a new battery-operated "smart bridge" is being developed to process high complexity data, like stream of images, and low complexity data, like temperature, on site. These features allow real-time monitoring and reduce the need of constant communication with a remote server, which may be accessible only with limited bandwidth due to the physical location of the monitoring system (e.g. mountains).

The Beesper SmartBridge system is a solar-powered and battery-operated intelligent monitoring device that acts as a Beesper bridge and is also able to perform advanced processing on site. The system must also be able to optimize power

F. Sassi (✉) · A. Bacchini
Camlin Italy s.r.l., Parma, PR, Italy
e-mail: f.sassi@camlintechnologies.com; a.bacchini@camlintechnologies.com

G. Massari
Politecnico di Milano, Milano, Italy
e-mail: giuseppe.massari@polimi.it

© Springer International Publishing AG, part of Springer Nature 2019
W. Fornaciari, D. Soudris (eds.), *Harnessing Performance Variability in Embedded and High-performance Many/Multi-core Platforms*, https://doi.org/10.1007/978-3-319-91962-1_13

consumption, in order to be deployed in remote areas with power supply provided by only solar power and battery.

The main applications developed are:

- *Target-based landslide movements monitoring*, using frequent images from cameras, and feature extraction, using machine vision and particle swarm optimization;
- *Markerless rockfall and rockwall composition identification*, using frequent images from cameras, through machine vision feature extraction and machine learning classification;
- *Dilatation of a rockwall analysis*, through machine learning using data from a wireless sensor network (WSN);

and two support applications as

- Network traffic manager
- Power monitor and board monitor logger

A specific installation of the system may be composed by only a subset of those applications, depending on the monitoring requirements; in particular, this chapter will focus on the latter two main applications.

The artificial intelligence system will be able to adapt itself to each landslide or rockwall condition, by performing the training of the system on the specific location, thus adapting the internal memory. This process will lead to better performance, plus lowering the cost of the solution and the time to market, with respect to a custom made system for each scenario.

The HARPA methodology allows to dynamically adjust the system power consumption, by changing applications' set of resource requirement (else said Application Working Mode or AWM), thus enabling to an improved quality of service, extended battery life and higher reliability.

13.1.1 Hardware Prototype

The Beesper SmartBridge system is composed by a solar panel, a solar charger, a DC/DC converter and a battery which provides power to a multicore ARM board. Internet connection can be provided by various means like UMTS/LTE, Ethernet, WiMAX or point-to-point radio connections.

Connection to the Beesper WSN is provided by a standard Beesper Bridge directly connected to the Beesper SmartBridge. The custom radio interface will be integrated in the Beesper SmartBridge in the final hardware version.

The system has been assembled in a weatherproof box with filtered air vents and it has been installed on a pole at about 3 m height to avoid tampering.

Details of the hardware components may be found in Table 13.1 and a picture can be found in the left part of Fig. 13.1.

Table 13.1 Beesper SmartBridge prototype hardware components

ARM board	Odroid-XU3
	multicore ARM platform based on the Samsung Exynos5422 with a Cortex A15 2.0 GHz quad core and a Cortex A7 quad core CPUs
Camera	USB uEye 4.92 Mpix colour
	(2560 × 1920 pixels) in an IP 30 casing
Battery	YUASA NP24-12I
	lead acid rechargeable battery (VRLA), 12 V, 24 Ah
DC/DC converter	Mean Well SD-50A-5 DC/DC
	Vin 12 V—Vout 5 V—Power 50 W
Solar panels	2 modules Photon PM020
	Max power 2 × 20 W = total 40 W
Solar charger	EPsolar Tracer1215BN
	12–24 V auto work—Max PV input power 130 W @ 12 V—RS485 communication

Fig. 13.1 *Left:* Beesper SmartBridge prototype. *Right:* Beesper SmartBridge prototype installed on a pole

13.1.2   Applications

The two core applications of the Beesper SmartBridge provide the main monitoring capabilities of the system and are developed to provide the state-of-the-art performance.

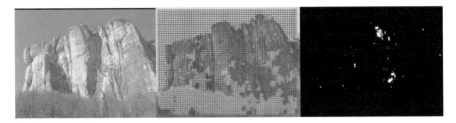

Fig. 13.2 Video Rockfall detection application. *Left:* Original image. *Centre:* Classification of vegetation, rock and sky. *Right:* Rockwall modification detection

13.1.2.1 Video Rockfall Detection

The aim of this application is to check if any rockfall, or major modification, has happened on the whole rock wall monitored using a high-resolution camera.

It is possible to identity four major processing steps:

1. Once the application is started, it will power up the camera and acquire a set of images in fast sequence and will evaluate the quality to select the best one for further processing. A sample image is shown in the left part of Fig. 13.2. In case of insufficient image quality (e.g. low light, heavy rain, . . .), the system may inhibit further application executions for a period of time.
2. Second step is to perform the actual computation by classifying the target image and tracking the relevant part of the image to detect rockfalls as shown in Fig. 13.2.
3. Third step is to save selected image and results to a temporary local drive storage, to queue for remote transmission. During the current test deploy phase, an additional local copy for long-term storage is written to a local flash memory.

After these three steps, the application waits to be rescheduled by HARPA-OS.

13.1.2.2 Extensometer Analysis

The aim of this application is to model, classify, and predict the behaviour of the rockwall boulders, monitored using a set of extensometers and temperature wireless sensors, to detect anomalies and dangerous events. The proprietary extensometers sensors used in this application are able to measure submillimetre rock displacement highlighting very subtle variation that may be correlated with the temperature of the rock. The site selected to host the HARPA application deployment is already monitored using a standard Beesper Wireless Sensor Network, thus a dataset composed of about 300 days is already available. The data is retrieved from the Beesper server using a set of API via the Internet. During the development phase, the whole historical dataset of the extensometers data has been downloaded and processed in order to train the machine learning system based on a proprietary framework based on hierarchical temporal-aware neural networks [1]. The historical data is classified by experts (geologists) between normal and not normal behaviour and then a model is trained for each extensometer. During run time, the updated data for each extensometer is retrieved from the Beesper server and is processed by the corresponding model. The model outputs the probability that the current situation represents a "not normal" behaviour or an anomaly.

13.1.3 Application Testing

The Video Rockfall detection application and the Fessurimeter analysis application have been tested to identify the power consumption sources on the ARM board, without the other hardware components of the Beesper SmartBridge, like solar charger or DC/DC converter. The test has been performed using the ODROID Smart Power device[1] which collects voltage, current and power of the system load. Moreover, this device, by cooperating with the on-board electronics, is able to measure the energy consumption for the USB, SoC, Ethernet and board. The values for the camera and other devices are derived from laboratory experiments.

Table 13.2 reports the time and energy needed for each applications' execution, while Table 13.3 reports the energy consumption per board component for each application execution computed in a controlled environment.

The two applications have a quite different power consumption profile and time requirements due to the specific processing performed, as it is possible to observe in Fig. 13.3.

Table 13.2 Time and energy comparison of the two main applications

	Video Rockfall detection application	Fessurimeter analysis application
Time		
Acquisition time (s)	11.4 (22%)	341.2 (90%)
Processing time (s)	39.8 (73 %)	38.7 (10%)
Total time (s)	51.2	379.9
Energy		
Acquisition energy (Wh)	0.0117 (13%)	0.2758 (79%)
Processing energy (Wh)	0.0752 (87%)	0.0727 (21%)
Total energy (Wh)	0.0869	0.3485

---

[1]The ODROID Smart Power is an easily deployable power supply that collects voltage, current and power of the system load to enable developers to optimize energy consumption. http://www.hardkernel.com/main/products/prdt_info.php?g_code=G137361754360.

Table 13.3 Energy consumption per board component for each application execution computed in a controlled environment of the two main applications

	Video Rockfall detection		Fessurimeter analysis	
Camera	5.46%	0.0047 Wh	6.06%	0.0211 Wh
USB	6.55%	0.0056 Wh	12.11%	0.0422 Wh
Board	32.73%	0.0284 Wh	60.55%	0.2110 Wh
SoC	33.55%	0.0291 Wh	7.76%	0.0270 Wh
Ethernet	4.91%	0.0042 Wh	9.08%	0.0316 Wh
Fan	12.72%	0.0110 Wh	3.08%	0.0107 Wh
Other	4.08%	0.0035 Wh	1.35%	0.0047 Wh

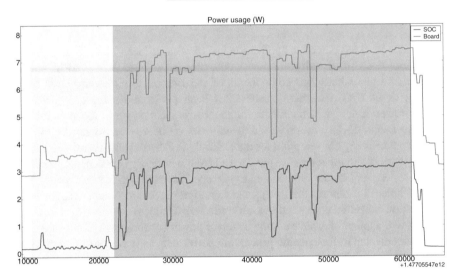

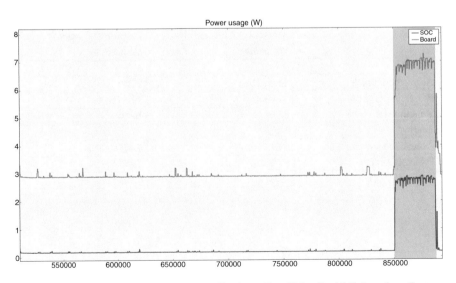

Fig. 13.3 Power usage of the two main applications. *Top:* Video Rockfall detection. *Bottom:* Fessurimeter analysis

The Video Rockfall detection application has a quicker acquisition phase where the connected camera captures a set of pictures and then the ARM board computes the results. Between different application executions, the system is idle.

The Fessurimeter analysis application has a much longer acquisition phase where the systems query and wait for the WSN data from a remote server via the Internet, thus the system is actually idle and the actual power consumption is not directly affected by this activity. This behaviour may lead to consider the second application more expensive in terms of total energy consumption per cycle (0.3438 Wh vs. 0.0834 Wh) while considering only the processing part it is less expensive (0.0685 Wh vs. 0.0718 Wh) than the camera-based one. In particular, analysing the distribution of the energy consumption per component it is possible to understand that most of the consumption in the Fessurimeter analysis application is due to the idle consumption of the board itself and not the SoC during the active processing of the data since it accounts about 20% of the overall energy consumption.

13.1.4 HARPA Integration

HARPA-OS provides a managed environment where to run the applications. HARPA-OS allocates resources to the applications depending on a set of policies, like temperature of the system, available energy, system load and data from the system monitors. The allocation is done by reconfiguring the application to a specific AWM (Application Working Mode) defined and profiled offline. The different AWM represent different application modality with different computational requirements. For instance, the application may reduce the accuracy of the algorithm or the duty cycle. Once the computational unit has finished, the applications must return the control to HARPA-OS in order to be, eventually, reconfigured. This interval should be in the seconds range. Given this workflow, the applications better suited to exploit HARPA-OS are the ones that analyse streaming data.

In the Beesper SmartBridge system, a specific logger reads current energy status, and associated resources from the memory, and provides the information to HARPA-OS. These "monitors" used by HARPA-OS are:

- Current battery level and
- Current power provided by the solar panels.

The "knob" used by HARPA-OS to tune the system behaviour is the applications duty cycle, in particular the different AWM are defined in Table 13.4. The duty cycle for the different applications depends also on the physical phenomena to be monitored and on the sensors used; for instance, the wireless extensometers provide a new value every 15 min.

HARPA-OS will reconfigure the applications to different AWM depending on the monitors values according to Table 13.5 in order to maximize the uptime of the system, by selecting a less-power hungry configuration when the energy is scarce, and to improve the quality of service, by selecting a higher monitoring frequency when energy is abundant.

Table 13.4 Application Working Mode (AWM) duty cycle

	Video Rockfall detection	Fessurimeter analysis
AWM 0	30 min	60 min
AWM 1	10 min	45 min
AWM 2	1 min	30 min
AWM 3	No delay and no waiting time in case of acquisition error	15 min

Table 13.5 AWM selection

Selected AWM		Power from solar panels (W)			
		$W <= 3$	$3 < W <= 20$	$20 < W <= 30$	$W > 30$
Battery charge status (%)	$\% <= 10$	0	0	1	1
	$10 < \% <= 40$	1	1	1	2
	$40 < \% <= 60$	1	1	2	3
	$60 < \% <= 80$	2	2	3	3
	$\% > 80$	3	3	3	3

13.2 Real-World HARPA Testing

The selected location to test the HARPA system is Pietra di Bismantova, a famous touristic location for climbers and excursionists in the district of Castelnovo ne' Monti (Reggio Emilia, Italy). This location is already monitored using a standard Beesper system, comprising cameras and WSN since January 2015. This system provides the internet connectivity via Ethernet to the Beesper SmartBridge only for installation convenience. The mains power is not connected to the Beesper SmartBridge.

The HARPA-based system has been installed in Bismantova from 13th November 2015 to 18th September 2016 and it is shown in Fig. 13.1. Due to development constraints, the system was initially deployed with only the Video Rockfall detection application while the Fessurimeter analysis application was deployed on 20th April 2016.

The total energy generated (and used) by the whole system during the whole testing period of 310 days is 25.89 kWh as calculated by the solar charger. After taking in account battery inefficiencies and solar charger's energy consumption, the energy spent by the computation system (including the camera) for the whole period is 19.6 kWh.

The idle energy required by the system is computed by considering all the logged data when no active process is running; the resulting average idle power is 6.0784 W.

The target is to reach the maximum uptime of the system with the higher quality of service possible. The system shuts off during night time since there would be no light to perform image-based analysis.

13.2.1 Collected Data Analysis for Each Application Working Mode

During the testing period, the system experienced all weather and illumination conditions which allowed to test system performance when the available energy was scarce or abundant; in particular, all the different AWM were selected during the test, thus it is possible to compute aggregate statistics over all the system AWM.

13.2.1.1 Load Power Analysis

Left part of Fig. 13.4 shows the aggregate load power statistics over the whole testing period. The height of the bin reflects the time spent by the system within the specific load power. Since the system is idle most of the time, it is possible to observe a peak around 6 W, which is the idle power consumption of the system. The right part of Fig. 13.4 shows a detail of the right tail of the previous plot. In this case, it is possible to observe a difference in the load power of the system in different AWM; in particular, AWM 2 and 3 highlight a considerable amount of time spent at high load power.

13.2.1.2 Solar Panel Power and Battery Level Analysis

Figure 13.5 shows the battery level and solar panel power distributions in the various AWM. It is possible to observe how the logged battery level and power generated from the panel correlates with the selected AWM of the system as expected. In

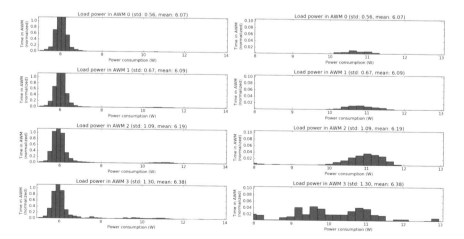

Fig. 13.4 *Left:* Power used by the system in each Application Working Mode (AWM). *Right:* Detailed view of the distribution "tail"

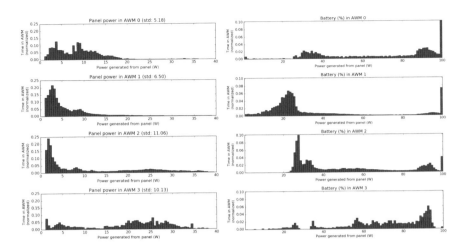

Fig. 13.5 Solar panel and battery percentage statistics in each AWM

particular, where the battery level and generated power are low the selected AWM is lower, while with high battery level or high generated power the selected AWM is higher.

It has been observed that the battery level is not reliable during active charging by the solar panels. In particular, it reaches 100% in a very short amount of time and the actual energy accumulated is not represented by the battery value. The current solution is a custom filtering to ignore non-realistic values while a more appropriate solution would require dedicated hardware components to evaluate the effective amount of charge flowing in the battery.

13.2.1.3 Application Working Mode Power Consumption Analysis

By further analysing the logged data, it is possible to calculate average energy for each run of each application AWM. Table 13.6 shows that the influence of the AWM can impact the energy requirement of the system up to an increase of 35%. A dynamic resource allocation system like HARPA-OS should be able to exploit this property.

As previously discussed, the average energy for the Fessurimeter analysis application is affected by the length of the acquisition phase where the system is idle; this behaviour shifts the average application energy to the idle one. Moreover, the very high consumption of the Video Rockfall detection in AWM 3 is also due to the removal of a wait period in case of low-quality image (e.g. low light, heavy rain, fog, bad camera initialization, ...).

Table 13.6 AWM average power consumption

	AWM	Average power (W)	Increase from idle power 6.0784 W
Video Rockfall detection	0	6.22	+2.3%
Execution time: 51.2 s	1	6.23	+2.4%
	2	7.08	+16.5%
	3	8.23	+35.3%
Fessurimeter analysis	0	6.05	–
Execution time:	1	6.10	–
38.8 s/379.9 s	2	6.14	+9.8%
	3	6.69	+10.0%

13.2.1.4 System Temperature Analysis

HARPA-OS provides an innovative scheduler, *Tempura*, able to allocate the available resources to the different applications, assuring that the thermal behaviour of the system is kept under control and the system performance is guaranteed.

During the testing period, the system has not reached critical temperatures leading to an emergency shutdown.

By further analysing the logged data, it is possible to compute a histogram representing the temperature distribution logged in the various AWM. Figure 13.6 highlights how the current AWM is directly linked to CPU temperature. In particular, it is possible to observe a maximum temperature reached by the system in all AWM; this behaviour will be analysed in Sect. 13.2.1.5.

Moreover, during the testing period the box temperature has never surpassed 30 °C and the CPU temperature in AWM 2 and 3 is not able to reach down to the enclosure box temperature since the duration of the Video Rockwall detection application is (or is very close to) the scheduling cycle, thus the system is always under active load.

13.2.1.5 Execution Time Analysis

The *Tempura* scheduler allocates the available resources in order to guarantee constant execution time. Figure 13.7 shows the correlation between CPU temperature (on the horizontal axis) and execution time (on the vertical axis) of the Video Rockwall detection application where the colour is the frequency of the event. It is possible to observe how the execution time clusters around two different values independently from the temperature value. This condition is generated by different exit conditions of the Video Rockfall detection application. If the image quality is not sufficient, the last stage of processing is skipped, thus resulting in a lower execution time.

This result indicates the benefits achieved with the HARPA-OS controlling the application scheduling with the Tempura resource allocation policy:

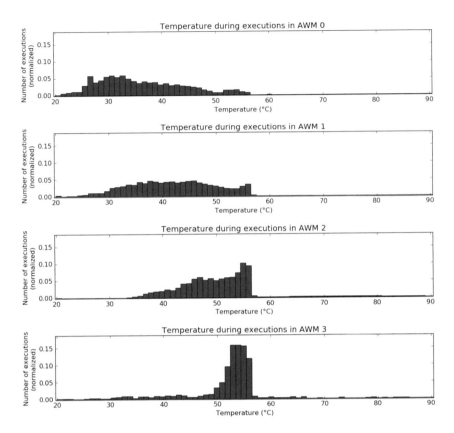

Fig. 13.6 System temperature statistics in each AWM

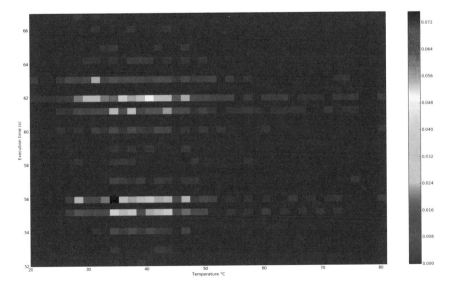

Fig. 13.7 Correlation between CPU temperature and execution time

- Execution time is independent from the temperature of the system;
- The maximum temperature of the system is under control and does not lead to safety shutdowns (thus no QoS issues): the hard threshold of 95 °C has never been reached even under heavy load of the two applications running at the same time in the highest AWM during summer time;
- When two applications are running, the resources are reallocated to guarantee a predictable performance level.

13.2.2 Performance Analysis

The main target of the Beesper SmartBridge system is to ensure a constant monitoring service of the rockwall, thus the main performance measure of is the uptime of the system. The uptime of the system in Table 13.7 has been computed by analysing the logs and by considering when the system was able to perform a successful operation (data acquisition, data processing and result communication) during a day.

Nevertheless, this coarse measure is not sufficient to evaluate the true performance of a monitoring system. To address this, two new measures of the quality of service of the system are introduced and will be described in the following sections:

- Uptime Quality of Service (UQoS)
- Performance Quality of Service (PQoS)

13.2.2.1 Uptime Quality of Service Analysis

The *Uptime Quality of Service* (UQoS) has been defined based on the percentage of time the system is active over 1 day. This index is directly correlated with the reliability of a monitoring system as can be defined as in Table 13.8. The colour

Table 13.7 Beesper SmartBridge uptime

Number of total days	310
Number of active days	286
Ratio	92%

Table 13.8 Uptime Quality of Service (UQoS) definition

	Uptime percentage	Status	Colour
UQoS-0	<10%	Insufficient UQoS	Red
UQoS-10	10% <> 50%	Minimum UQoS	Orange
UQoS-50	50% <> 90%	Good UQoS	Yellow
UQoS-90	>90%	Optimal UQoS	Green

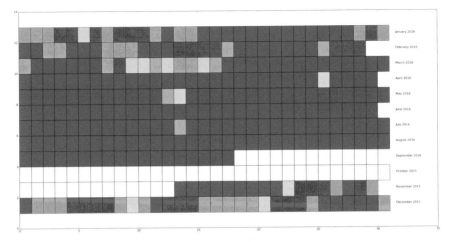

Fig. 13.8 Daily Uptime Quality of Service (UQoS) of the HARPA system over the calendar months

indicated in the table will be used in the figures containing the Beesper SmartBridge UQoS results.

Figure 13.8 shows this index as a colour-coded scale on a calendar in order to highlight the time of the year with respect to the performance of the system. The colour of the days represents the UQoS that the system achieved on the specific day.

To perform this analysis, the applications' current AWM has not been considered since all applications AWM assure a minimum level of service.

It is possible to observe in Fig. 13.8 how during spring and summertime the system is active with the best possible UQoS almost the whole time thanks to the available increased solar power. The only red day in May 2016 is due to a software bug that prevented the system to boot up, condition that was automatically solved the day after without requiring an on-site intervention.

Table 13.9 and Fig. 13.9 show the percentage of time at which the system reached each UQoS month by month. It is possible to observe that the system can reach the minimum UQoS-10 for most of the time during the whole year and the optimal UQoS-90 during late spring, summer and early autumn. From an application point of view, these results indicate the suitability of these performances to create a battery-powered monitoring system able to provide consistent performance for most part of the year.

These results will be compared in Sect. 13.3.2.2 with the one obtained simulating the system performance without HARPA-OS.

Table 13.9 Percentage of time the system reached each UQoS month by month

Month	UQoS-10	UQoS-50	UQoS-90
11/15	**76%**	65%	59%
12/15	**68%**	29%	23%
1/16	**77%**	48%	45%
2/16	**76%**	52%	52%
3/16	**97%**	77%	65%
4/16	100%	100%	**97%**
5/16	97%	94%	**90%**
6/16	100%	100%	**100%**
7/16	100%	97%	**97%**
8/16	100%	100%	**100%**
9/16	100%	100%	**100%**

In bold, during winter months the system reached sufficient UQoS for the majority of the time while in spring and summer the system reached optimal UQoS for the majority of the time.

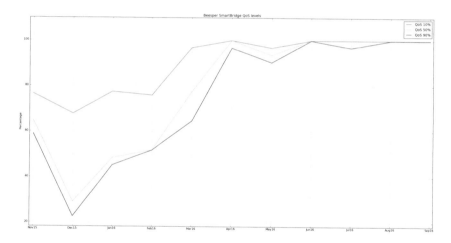

Fig. 13.9 Percentage of time the system reached each UQoS month by month

13.2.2.2 Performance Quality of Service Analysis

The *Performance Quality of Service* (PQoS) measures how often the Beesper SmartBridge monitors the environment; thus, it is able to provide a measure of the system performance in assessing the rockwall condition.

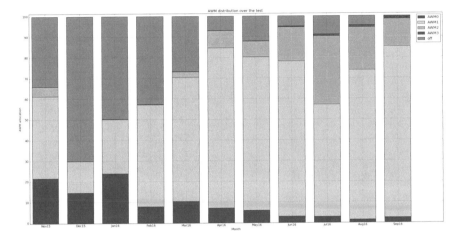

Fig. 13.10 PQoS of the system: percentage of time for each month for each AWM

This index is directly correlated with the active AWM of the system as:

- AWM 0 = very low PQoS
- AWM 1 = sufficient PQoS
- AWM 2 = good PQoS
- AWM 3 = best PQoS

A further analysis of the log files allows to understand the AWM distribution over the different months in order to understand the impact of the HARPA-OS scheduler. In particular, Fig. 13.10 shows how the AWM 0 was more present when less solar power was available while AWM 2 and 3, that allow a higher quality of service, can be safely scheduled when there is no risk to reduce the battery life. Table 13.10 contains the detailed data of Fig. 13.10 and highlights how the AWM 0 and AWM 2 are active for a consistent amount of time during the months when power available was, respectively, scarce or plentiful. The data in these plots and table are computed on the overall time of the month, not on a day basis as the UQoS.

These results are consistent with the data presented in Fig. 13.4, where the various AWM show different power requirements.

These strong results and the one from the UQoS analysis show the effectiveness of the HARPA methodology to increase the monitoring performance of the Beesper SmartBridge without impacting the uptime of the system when a greater amount of energy is available.

In order to understand the impact on the system uptime also during the other periods, a comparison with a simulated system is presented in Sect. 13.3.

Table 13.10 PQoS of the system: percentage of time for each month for each AWM

	AWM 0	AWM 1	AWM 2	AWM 3	On total	Off + Err
Nov15	**22%**	40%	5%	0%	66%	34%
Dec15	**15%**	15%	0%	0%	30%	70%
Jan16	**24%**	26%	0%	0%	50%	50%
Feb16	8%	49%	0%	0%	57%	43%
Mar16	10%	60%	3%	0%	73%	27%
Apr16	7%	78%	**8%**	0%	93%	7%
May16	6%	74%	**8%**	0%	88%	12%
June16	3%	75%	**16%**	0%	95%	5%
July16	3%	54%	**33%**	1%	91%	9%
Aug16	1%	73%	**21%**	1%	95%	5%
Sept16	2%	83%	13%	1%	100%	0%

In bold the months where the impact of dynamic AWM scheduling is more relevant

13.3   Evaluating HARPA Impact: A Simulation Approach

HARPA-OS provides a dynamic resource allocation system thus a variable level
of performance and power consumption. A simulated system with standard static
resource allocation is used to evaluate HARPA-OS impact on the Beesper Smart-
Bridge uptime.

The simulated system exploits the data logged by the Beesper SmartBridge
during the testing period, in particular the actual power consumption and available
solar radiation in the installation site. To further increase the accuracy of the
simulation, a set of laboratory tests were performed to determine the efficiency of
the various Beesper SmartBridge components, in particular the battery, the DC–DC
converter and the solar charger. All these information have been used to create a
custom energy simulator of the Beesper SmartBridge.

13.3.1   Available Energy Simulation

The total solar power generated is autonomously logged by the solar charger device,
to which the system queries every 30 s to save the current value. Since the system
was not active the whole time, thus after a period of time when the system was off,
the first value read from the solar charger corresponds to the total amount of energy
generated since the last query.

To obtain a reliable simulation model, it is important to have realistic solar
radiation data during each hour, and not only the total amount. Thus, it is necessary
to perform an interpolation process to estimate the energy generated while the
system was off while keeping the total energy generated consistent.

The total amount of energy generated by the Beesper SmartBridge during the testing period has been measured as 25.89 kWh; while the total amount of energy available to the simulated system has been computed as 25.48 kWh exposing a difference of −1.58% which allows to compare the simulated values with the real system. These values do not account for the solar charger power consumption thus have been scaled accordingly before the usage in the simulator.

13.3.2 Application Working Mode Simulation

Two different simulation tests are performed considering the system running always in AWM 1 (sufficient PQoS) and AWM 3 (best PQoS), as defined in Sect. 13.2.2.2.

Moreover, as seen in Fig. 13.10, AWM 1 is the most active AWM; thus, it is likely to expect similar results regarding the uptime to the HARPA-based system, beside the exploitation of AWM 0 to save power. Instead, the simulation results with AWM 3 are likely to be quite different given the much higher power requirements.

The simulation is performed by creating a model of the system that receives as input the simulated available energy and simulates application executions, battery and solar panel status hour-by-hour. The simulated test has the same temporal duration as the real-world test in order to compare them. The required energy for each application execution is computed using the data collected during system testing depending on the selected AWM.

In order to have a fair comparison between the real system and the simulated one, it has been considered only the power consumption related to the Video Rockwall detection application since it is the only application that was active during the whole recording period.

13.3.2.1 Uptime Days Comparison

Table 13.11 reports the uptime results of the simulated systems compared to the real HARPA-based system and the energy requirements for AWM 1 and AWM 3. The uptime is computed, as in Sect. 13.2.2, considering when the system performs a successful operation (data acquisition, data processing and result communication) during a single day. From this table, it is possible to observe how the HARPA system achieves higher uptime values, even accounting for the energy error of 1.58% previously reported. Moreover, the small difference between the HARPA system and the simulated systems is consistent with the expectation, considering how the uptime is computed.

Nevertheless, this very high-level comparison is not able to highlight short-term differences between the different approaches, thus a more fine-grained measure, like the UQoS of the simulated systems, should be computed.

Table 13.11 Uptime comparison between the simulated and HARPA-based systems

	Energy req.	Uptime days/310		Days	Uptime %
HARPA	Variable	286	92.26%		
AWM 1	6.2250 Wh	267	86.13%	−19	−6.13%
AWM 3	8.2251 Wh	246	79.35%	−40	−12.90%

Table 13.12 UQoS results of the simulated systems compared to the HARPA-based system

	Colour	HARPA	AWM 1		AWM 3	
UQoS-0 Insufficient	Red	30	48	60.00%	79	163.00%
UQoS-10 Minimum	Orange	38	20	−47.00%	43	13.00%
UQoS-50 Good	Yellow	10	10	0.00%	49	390.00%
UQoS-90 Optimal	Green	232	232	0.00%	139	−40.00%

13.3.2.2 Uptime Quality of Service Comparison

In Sect. 13.2.2.1, the *Uptime Quality of Service* (UQoS) has been defined as the percentage of time the system is active over 1 day and the results over the HARPA-based system were presented. This index is directly correlated with the reliability of a monitoring system as can be defined as in Table 13.8. The colour indicated in the table will be used in the figures containing the Beesper SmartBridge UQoS results.

Table 13.12 reports a quantitative analysis and it is possible to observe the HARPA system compared to the simulated systems.

With respect to the AWM 1 simulated system, the HARPA system has a reduced number of UQoS-0 days, where the minimum UQoS is not reached, by increasing the number of UQoS-10 days thanks to its dynamic scheduling of AWMs. This is a strong evidence of the effectiveness of a dynamic approach, such as the one presented in HARPA-OS, that allows to reach a minimum level of quality of service by enforcing dynamic energy saving measures.

The comparison with the simulated AWM 3 system shows, as expected, a greatly reduced UQoS; in particular, it is not able to reach very good UQoS-90 level consistently.

As a qualitative comparison between the HARPA system and the simulated systems, it is possible to analyse the UQoS in a calendar plot as shown in Fig. 13.11. The colour of the days represents the UQoS that the system achieved on the specific day. A similar plot was presented for the HARPA system in Fig. 13.8.

The comparison between the HARPA system and the AWM 1 simulated system shows a great similarity of UQoS level, as expected, during most of the year. Nevertheless, it is important to remember that the HARPA system provides a higher

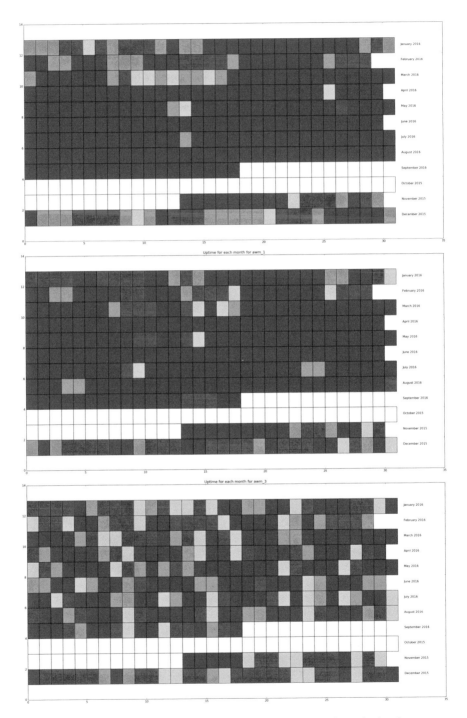

Fig. 13.11 UQoS comparison between the real and simulated systems in a calendar plot

Table 13.13 UQoS comparison between the real and simulated systems month by month

Month	#Days	AWM 1 UQoS (%)			AWM 3 UQoS (%)			HARPA UQoS (%)		
		10	50	90	10	50	90	10	50	90
11	17	88	75	69	65	65	35	76	65	59
12	31	52	35	29	42	32	13	68	29	23
1	31	68	55	52	61	48	29	77	48	45
2	29	72	62	55	62	48	31	76	52	52
3	31	87	81	74	81	68	55	97	77	65
4	30	100	**100**	**100**	83	**70**	**57**	100	**100**	**97**
5	31	90	**90**	87	84	**71**	**58**	97	**94**	**90**
6	30	100	**100**	**100**	93	**63**	**57**	100	**100**	**100**
7	31	97	**90**	87	90	**74**	**52**	100	**97**	**97**
8	31	97	**90**	**90**	81	**68**	**55**	100	**100**	**100**
9	18	83	**83**	**83**	72	**61**	**50**	100	**100**	**100**

level of performance, through AWM 2 and AWM 3, in particular during the months with more solar power available as demonstrated in Sect. 13.2.2.2.

Moreover, even during summer time, when the solar power is greatly available, a dynamic approach like HARPA has an impact on the UQoS of the system. The simulated AWM 3 system is not able to reach constant high UQoS while the AWM 1 system shows more UQoS-0 and UQoS-10 days than the HARPA system.

Table 13.13 reports the analytical data of Fig. 13.11 and highlights in bold how the simulated AWM 1 and HARPA systems show consistent results while the simulated AWM 3 system is not able to reach similar performance levels.

Figure 13.12 shows a qualitative comparison of the simulated systems and the HARPA system over the UQoS level reached during each month.

13.4   Conclusions

The Beesper SmartBridge prototype presented in this chapter is the demonstrator in the low-end of embedded system domain for HARPA.

The Beesper SmartBridge is a HARPA-based solar-powered and battery-operated intelligent monitoring device able to autonomously perform markerless rockfall and rockwall composition identification (using frequent images from cameras analysed through machine vision feature extraction and machine learning classification) and dilatation of a rockwall analysis (through machine learning using data from a WSN) as detailed in Sect. 13.1. Since the monitoring system is battery operated, the main target is to provide consistent monitoring performance while maximizing the system uptime.

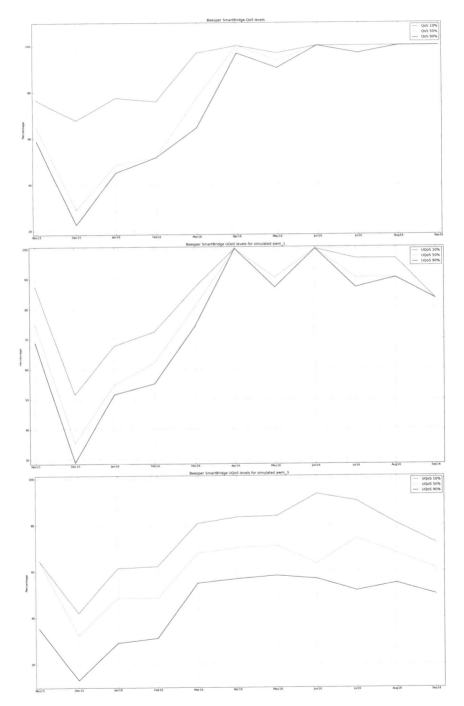

Fig. 13.12 UQoS comparison between the real and simulated systems month by month

The Beesper SmartBridge has been successfully installed at Pietra di Bismantova (Reggio Emilia, Italy) from 13th November 2015 to 18th September 2016 for a total of 310 days as reported in Sect. 13.2.

The laboratory testing of the system and applications, reported in Sect. 13.1.4, shows the suitability of the Beesper SmartBridge to the HARPA approach given that the various applications run concurrently with tunable power consumption.

The system was able to log accurate information during the whole test duration and Sect. 13.2.1 presents an analysis about the load power, battery percentage and power generated from the solar panel with respect to the various AWM that highlights how the AWM selection process based on external factors is successful.

Finally, the average energy required by each application AWM is computed from the logged data giving a real-world estimation of the energy need of this system. These results, reported in Table 13.6, show that the selection of an AWM has a significant impact on the total energy requirements (up to 35% more than the idle power), thus a dynamic resource allocation approach like HARPA-OS is appropriate.

HARPA-OS provides a temperature and energy aware scheduler, *Tempura*, and Sect. 13.2.1.5 reports its impact during the whole test period. It has been demonstrated how the *Tempura* scheduler allocates resources to the two applications in order to provide consistent execution time, while keeping under control the temperature of the system. It is worth remarking how the system could operate correctly both during winter time, with low temperatures, and summer time, with higher environmental temperature.

The Beesper SmartBridge is a monitoring system where the various applications duty cycle directly influence the quality of the performance of the system; thus, the duty cycle is the main "knob" to tune system performance. The various *Application Working Mode* (AWM) are defined with different "wait intervals" between execution, thus exhibit different level of average power consumption. The main system "monitors" are the battery level, available solar power and system temperature.

To evaluate the performance of the system, three different measures of the quality of service have been defined in Sect. 13.2.2 and two simulated systems with static AWM definition have been created.

The simulated system exploits the data logged by the Beesper SmartBridge during the testing period, in particular the actual power consumption and available solar radiation in the installation site. To increase further the accuracy of the simulation, a set of laboratory tests were performed to determine the efficiency of the various Beesper SmartBridge components, in particular the battery, the DC–DC converter and the solar charger. All these information have been used to create a custom simulator.

The *Uptime days* performance evaluation resulted in the HARPA-OS-based system with an uptime of 286 over 310 days (92.26%) while the simulated systems in AWM 1 and AWM 3 scored, respectively, 267 (86.13%) and 246 (79.35%) days of uptime. The small difference between the HARPA system and the simulated systems is consistent with the uptime definition and the fact that AWM 1 is the most selected AWM by HARPA-OS as reported in Table 13.10.

The *Performance Quality of Service* (PQoS) focuses on the distribution of the application AWM during each month, from AWM 0 (very low PQoS) to AWM 3 (best PQoS) and shows in Sect. 13.2.2.2 the impact of the HARPA-OS dynamic scheduler. In particular, the AWM 0 is more present (up to 23.90% in January) when less solar power was available, while AWM 2 can be safely scheduled when there is no risk to reduce the battery life (up to 33.15% in July).

The results are confirmed when comparing the *Uptime Quality of Service* (UQoS) performance index as reported in Table 13.12 where it is possible to observe how the HARPA system, compared to the simulated AWM 1 system, can reduce the number of days with insufficient UQoS-0 of 60% by increasing the number of days with sufficient UQoS-10 of 47% thanks to its dynamic scheduling of resources.

From Table 13.13, it is possible to observe that HARPA-OS has also a strong impact during the months with more solar power available. For instance, the simulated AWM 3 system reaches only 58% of the time the UQoS-90, while the HARPA system can reach the same UQoS-90 almost 100% of the time during summer months.

It is not possible to compare the *Performance Quality of Service* PQoS performance index since the AWM is fixed in the simulated systems.

These important results demonstrate how HARPA-OS is able to allow a greater PQoS when possible, while keeping a minimum PQoS and consistent UQOS during the rest of the time.

All these results provide strong evidence of the effectiveness of a dynamic approach, such as the one presented in HARPA-OS, in increasing the overall quality of the service provided by a real-world system equipped with such functionality. Moreover, the Beesper SmartBridge, with HARPA-OS, has been installed for almost 1 year outdoor, without any need of maintenance, demonstrating the stability and reliability of the hardware and software solutions and providing a good level of performance. Finally, the system has collected an extremely valuable dataset of high-quality pictures, temperatures and rock dilatations during almost one full year that will be used to greatly improve the quality of the computation results in the next development phase.

Reference

1. WO2017012677 - system and computer-based method for simulating a human-like control behaviour in an environmental context. Application number: WO2015EP66952, Filing date: 2015-07-23

Index

© Springer International Publishing AG, part of Springer Nature 2019   319
W. Fornaciari, D. Soudris (eds.), *Harnessing Performance Variability
in Embedded and High-performance Many/Multi-core Platforms,*
https://doi.org/10.1007/978-3-319-91962-1

Printed in the United States
By Bookmasters